The Ecosystem Approach to Marine Planning and Management

The Ecosystem Approach to Marine Planning and Management

Edited by Sue Kidd, Andy Plater and Chris Frid

publishing for a sustainable future

London • Washington, DC

First published in 2011 by Earthscan

Earthscan Ltd, Dunstan House, 14a St Cross Street, London EC1N 8XA, UK

Earthscan LLC,1616 P Street, NW, Washington, DC 20036, USA

Earthscan publishes in association with the International Institute for Environment and Development

For more information on Earthscan publications, see www.earthscan.co.uk or write to earthinfo@earthscan.co.uk

ISBN: 978-1-84971 -182-1 hardback

ISBN: 978-1-84971-183-8 paperback

Typeset by Saxon Graphics Ltd, Derby

Cover design by Susanne Harris

A catalogue record for this book is available from the British Library

Library of Congress Cataloging-in-Publication Data

The ecosystem approach to marine planning and management / edited by Sue Kidd, Andy Plater, and Chris Frid.

   p. cm.

Includes bibliographical references and index.

ISBN 978-1-84971-182-1 (hardback) — ISBN 978-1-84971-183-8 (pbk.) 1. Marine ecology. 2. Marine pollution. 3. Maritime law. 4. Environmental law. I. Kidd, Sue. II. Plater, Andy. III. Frid, Chris.

QH541.5.S3E267 2010

333.95'616—dc22   2010037831

At Earthscan we strive to minimize our environmental impacts and carbon footprint through reducing waste, recycling and offsetting our CO_2 emissions, including those created through publication of this book. For more details of our environmental policy, see www.earthscan.co.uk.

Printed and bound in the UK by MPG Books, an ISO 14001 accredited company. The paper used is FSC certified.

MIX
Paper from
responsible sources
FSC
www.fsc.org
FSC® C018575

Contents

List of Figures, Tables and Boxes

Figures

Tables

Boxes

List of Contributors

Rhoda Ballinger is Lecturer in the School of Earth and Ocean Sciences, Cardiff University, UK

Adam Barker is Lecturer in Spatial Planning (Environment and Landscape), School of Environment and Development, University of Manchester, UK

Tom Barker is a limnologist and freshwater ecologist in the School of Environmental Sciences, University of Liverpool, UK

Jim Claydon was recently President of the Royal Town Planning Institute and is a consultant in Town Planning

Robert Duck is Professor of Environmental Geoscience in the School of Social and Environmental Sciences, University of Dundee, UK

Geraint Ellis is Senior Lecturer in the School of Planning, Architecture and Civil Engineering (SPACE) and Institute of Spatial and Environmental Planning, Queen's University, Belfast, UK

Chris Frid is Professor in Marine Biology in the School of Environmental Sciences, University of Liverpool, UK

Gillian Glegg is Associate Professor (Senior Lecturer) in Marine Management, School of Marine Science and Engineering, University of Plymouth, UK

Sture Hansson is Professor of Aquatic Ecology in the Department of Systems Ecology, Stockholm University, Sweden

Sue Kidd is a chartered town planner and Senior Lecturer in the School of Environmental Sciences, University of Liverpool, UK

Manos Koutrakis is a Senior Researcher (biologist-ichthyologist) at the Hellenic Fisheries Research Institute (National Agricultural Research Foundation), Kavala, Greece

Kirsty Lindenbaum is Maritime Planning Officer, Countryside Council for Wales (CCW), UK

Greg Lloyd is Head of School of the Built Environment, University of Ulster, UK

Chris Lumb is Marine Delivery Leader NW, Natural England, Kendal, Cumbria, UK

Ed Maltby is Professor of Wetland and Water Science in the School of Environmental Sciences, University of Liverpool, UK

Stephen Mangi is an Environmental Economist at Plymouth Marine Laboratory, UK

Ivona Marasovic is the Director of the Institute of Oceanography and Fisheries, Split, Croatia

Charlotte Marshall is studying for a PhD in the School of Marine Science and Engineering, University of Plymouth, UK

Tim Norman is Senior Manager, Planning, with the Crown Estate, UK

Temel Oguz is Professor of Physical Oceanography, Institute of Marine Sciences, Middle East Technical University, Mersin, Turkey

Frances Peckett is studying for a PhD in the School of Marine Science and Engineering, University of Plymouth, UK

Andy Plater is Professor in Physical Geography in the School of Environmental Sciences, University of Liverpool, UK

Sian Rees is studying for a PhD in the School of Marine Science and Engineering, University of Plymouth, UK

Jake Rice is National Senior Advisor – Ecosystem Sciences, for the Department of Fisheries and Oceans (DFO), Canada

Lesley Rickards is the Director of the Permanent Service for Mean Sea Level (PSMSL), National Oceanography Centre, Liverpool, UK

Leonie Robinson is a Lecturer in Marine Biology in the School of Environmental Sciences, University of Liverpool, UK

Lynda Rodwell is Lecturer in Ecological Economics, School of Marine Science and Engineering, University of Plymouth, UK

Stuart Rogers is Divisional Director, Environment and Ecosystems, Centre for Environment, Fisheries & Aquaculture Science (CEFAS), Lowestoft, UK

Hance Smith is a Reader in the School of Earth and Ocean Sciences, Cardiff University, UK

Tim Stojanovic is Teaching Fellow in Sustainable Development, School of Geography and Geosciences, Sustainability Institute and Scottish Oceans Institute, University of St. Andrews, UK

David Tudor is Marine Policy Manager with the Crown Estate, UK

Nedo Vrgoč is a Senior Researcher at the Institute of Oceanography and Fisheries, Split, Croatia

Nigel Watson is Lecturer in Resource and Environmental Management, The Lancaster Environment Centre, University of Lancaster, UK

List of Acronyms

BGS	British Geological Survey
BODC	British Oceanographic Data Centre
CalCOFI	California Cooperative Oceanic Fisheries Investigations
CAP	Common Agricultural Policy
CBA	Cost Benefit Analysis
CBD	Convention on Biological Diversity
CCAMLR	Convention on the Conservation of Antarctic Marine Living Resources
CEC	Commission of the European Communities
CEFAS	Centre for Environment, Fisheries and Aquaculture Services (UK)
CEM	Commission on Ecosystem Management
CFP	Common Fisheries Policy
COP	Conference of the Parties
CPR	continuous plankton recorder
DASSH	Data Archive for Seabed Species and Habitats
DFO	Department of Fisheries and Oceans (Canada)
DG	Directorate-General
DMS	dimethyl sulphide
DSS	Decision Support System
DST	decision support tool
EA	ecosystem approach
EBSA	ecologically and biologically significant area
EBSS	ecologically and biologically significant species
EC	European Commission
EEC	European Economic Community
EEZ	Exclusive Economic Zone
EOAR	Ecosystem Overview and Assessment Report
ERAEF	Ecological Risk Analysis for Effects of Fishing
ERDF	European Regional Development Fund
ERSEM	European Regional Seas Ecosystem Model
ESRC	Economic and Social Research Council
EU	European Union
EwE	Ecopath with Ecosim
FAO	Food and Agriculture Organization (UN)
FPZ	Fisheries Protection Zone
GDP	Gross Domestic Product
GES	Good Ecological Status

GFCM	General Fisheries Commission for the Mediterranean
GIS	geographic information system
HCR	Harvest Control Rule
HELCOM	Helsinki Commission
ICCAT	International Commission for the Conservation of Atlantic Tunas
ICES	International Council for the Exploration of the Sea
ICSU	International Council for Science
ICZM	Integrated Coastal Zone Management
IGO	intergovernmental organization
IOC	Intergovernmental Oceanographic Commission
IPC	Infrastructure Planning Commission
ITQ	individual transferable quota
ITR	Individual Transferable Right
IUCN	International Union for Conservation of Nature
IUU	illegal, unreported and unregulated
JNCC	Joint Nature Conservation Committee (UK)
LME	Large Marine Ecosystem
LOMA	Large Ocean Management Area
MaRS	Marine Resource System
MCA	multicriteria analysis
MCZ	Marine Conservation Zone
MEA	Millennium Ecosystem Assessment
MEDIN	Marine Environmental Data and Information Network
MERMAN	Marine Environment Monitoring and Assessment National
MHM	marine habitat mapping
MMO	Marine Management Organization
MP	management procedure
MPA	Marine Protected Area
MPS	marine policy statement
MSE	management strategy evaluation
MSFD	Marine Strategy Framework Directive
MSP	marine spatial planning
MSY	maximum sustainable yield
NERC	Natural Environment Research Council (UK)
NPS	national policy statement
NSIP	nationally significant infrastructure project
OM	operating model
PCB	polychlorinated biphenyl
RNLI	Royal National Lifeboat Institution
ROPME	Regional Organization for the Protection of the Marine Environment

SA	Sustainability Appraisal
SBSTTA	Subsidiary Body on Science, Technology and Technical Advice
SDM	species distribution model
SME	small to medium enterprise
SSB	spawning-stock biomass
TAC	Total Allowable Catch
TL	trophic level
UKHO	UK Hydrographic Office
UKOOA	UK Offshore Operators Association
UN	United Nations
UNCLOS	United Nations Convention on the Law of the Sea
UNEP	United Nations Environment Programme
UNESCO	United Nations Educational, Scientific and Cultural Organization
VMS	vessel monitoring system

Preface

The marine environment covers two-thirds of our planet's surface and is generally regarded as one of our most precious natural resources. It provides a wide range of essential goods and services upon which human well-being and, ultimately, all life on Earth depends. Since prehistoric times, humans have looked to the sea for food, transport, waste disposal, and cultural and spiritual fulfilment; more recently, coastal areas have become valued settings for recreation and tourism. Increasingly, the marine environment is being relied upon to underpin economic development through the provision of aggregates, fossil fuels, renewable energy, and so on. Alongside these very tangible benefits, modern science is revealing the reality of less tangible but in many ways more profound dependencies on the sea, including the vital role it plays in climate regulation and carbon capture. As appreciation of the intricate web of relationships between humans and the marine environment continues to grow, so too does our understanding of the complexity of impacts that human activity is having on the dynamics of marine ecosystems. In many ocean areas, there is a concern that the nature and level of human pressure, both direct and indirect, is unsustainable and is leading to significant deterioration of the health of our seas. This puts at risk the resources we need and the 'life support' system of our planet.

It is within this context that calls for improved planning and management of the marine environment, and specifically marine ecosystems, have emerged. For example, the United Nations Educational, Scientific and Cultural Organization (UNESCO) is promoting the widespread introduction of marine spatial planning, and the European Union (EU) is developing a new Integrated Maritime Policy designed to stimulate marine-related economic growth but at the same time protect the marine resource base. In the UK, the 2009 Marine and Coastal Access Act draws upon the experience of town and country planning and brings into being an integrated system of marine planning for the first time. Similar initiatives are now beginning to be put in place in many other parts of the world. A common feature in all these developments is an appreciation that more integrated and holistic forms of planning and management are required for our seas and that new arrangements must draw together the very best understanding of both the natural and human aspects of marine systems to find a more sustainable way forward.

This holistic perspective is at the heart of the ecosystem approach (EA) to natural resource planning and management that has been promoted over the past few years in both terrestrial and marine contexts. This book brings together expertise from natural scientists, social scientists and marine planning and management practitioners to promote a broader understanding of issues that need to be addressed in applying EA to the seas. It draws upon a series of Economic and Social Research Council and the Natural Environment Research Council-funded transdisciplinary seminars that were held at different institutions across the UK between 2007 and 2009. This book is a key output from the seminar series and seeks to disseminate the richness of discussions to a wider audience in order to:

- provide fresh insights into the operationalization of EA and related integrated methodologies for better planning and management of the marine environment;
- demonstrate the useful linkages between natural science and social science understandings and how they can inform and facilitate marine planning and management practice; and
- serve as a platform for the development of new transdisciplinary initiatives in this area.

The book is aimed at undergraduate and postgraduate students, practitioners and researchers coming from a range of disciplinary backgrounds who have an interest in EA to natural resource management and its application in the marine environment. It seeks to complement other excellent texts in the field. For example, *Oceans Past* edited by Starkey et al (2007) provides a stimulating historical perspective on human interactions with marine ecosystems with particular reference to fisheries and is a useful starting point for those who are new to the field, as well as a valuable reminder to old hands of why improved control over human activity in the sea is needed. *Ecosystem-based Management for the Oceans* edited by McLeod and Leslie (2009) is a valuable text which very much parallels this one with a special focus on experience in northern and central America. A detailed and timely review of the theory and international practice of zoning that is emerging as a key tool in marine planning and management is provided by Agardy (2010) in *Ocean Zoning: Making Marine Management More Effective*. Readers may also find *Managing Britain's Marine and Coastal Resources* edited by Potts and Smith (2005) a useful exploration of how one country with a strong maritime tradition is trying to deal more effectively with the multifaceted interactions between terrestrial and marine areas.

This volume attempts to contribute to the wider body of literature in two ways. The first is through its transdisciplinary outlook, with each of the chapters reflecting the joint effort of authors from different disciplinary backgrounds including inputs from practitioners. This broad understanding is the key to the delivery of EA and we hope that the book encourages others to step across disciplinary divides to engage with concepts and material that may be new to them in the search for more rounded analysis, fresh insights and the development of more robust ways forward. The second is through its European perspective. While the book deals with a subject matter that is of global relevance and draws on international understanding, the authors are mainly (although not exclusively) from Europe and much of the research and practice experience that underpins the book reflects this.

Chapter 1 sets the scene for following contributions by: explaining the origins, definitions and principles of EA and associated United Nations (UN) operational guidance; reviewing some of the lessons that can be drawn from existing experience of applying EA in non-marine and marine areas; and providing a transdisciplinary discussion of key issues that require further attention in developing EA in future planning and management of the sea. These include: developing the human dimension of EA; addressing key information challenges; and connecting marine planning and management understanding and concerns to wider agendas including integration with other planning activities and underlying debates about human development trajectories. These issues are by no means unique to the application of EA in marine ecosystems but they are intensified in the marine context for various reasons.

Chapters 2 and 3 relate to the first of these issues and illustrate why understanding of and engagement with the human dimension of marine systems is so critical to effective planning and management of the sea. The first of these chapters looks at the commonalities between planning for the land and planning for the sea including their shared concern to control human use of and impacts on natural resources in accordance with wider social, economic and environmental objectives. It provides examples of how terrestrial planning experience can inform debate about the purpose and process of planning for marine areas. Chapter 3 reviews the development of EU marine and maritime policy over the first part of the 21st century and illustrates the strong link between land-based concerns and marine planning agendas. It highlights the difficulties of combining political priorities linked to job creation and continuing economic growth with sustainable use of marine resources and how European countries are trying to deal with these issues with particular reference to the Common Fisheries Policy (CFP) and the directives concerned with the management of the coastal and marine environment. These developments are set against the background of constitutional changes, including EU enlargement, national devolution on the one hand, and EU development related to the Maastricht, Nice and Lisbon treaties on the other.

Chapters 4 and 5 focus their attention on key information challenges in marine planning and management. Chapter 4 considers the concept of ecosystem goods and services which has emerged as a key tool for delivery of EA and its value in the marine context. It explores: the scale and value of marine ecosystem goods and services in the UK; the role of the marine ecosystem processes in delivering these valuable goods and services; the major pressures on these processes; and possible planning and management responses to these pressures. Chapter 5 draws upon case study experience in a range of countries to provide an overview of data and modelling tools with which the scientific evidence base is able to support adaptive planning and management which are central to EA. The chapter highlights the challenges involved in utilizing the available evidence base, tools for assessment and the current framework for developing shared knowledge.

Chapter 6 concludes with a summary of the main findings that emerge from the earlier chapters and also with a discussion of overall conclusions from the seminar series related to natural science, social science and policy and governance perspectives. The chapter closes with the identification of future research priorities that could assist the more effective delivery of EA in marine planning and management activities over the next few years.

Before turning to these chapters we would like to offer our thanks to all those who have made this book possible. Firstly we must thank the Economic and Social Research Council (ESRC) and the National Environment Research Council (NERC) for supporting the transdisciplinary seminar series upon which the book is based. This provided a key opportunity for a wide range of academics and practitioners with an interest in developing new approaches to marine planning and management to come together and the chapters presented here have benefited greatly from the input of all participants. We must also thank the University of Liverpool, the National Oceanography Centre, Queen's University Belfast, Cardiff University, the University of Dundee and the University of Plymouth for hosting the seminars, and all the people, especially Emma Walsh, who made these seminars possible and helped them to run so smoothly. Many individuals have contributed at various points in time to the production of the chapters and you will see that the list of authors is quite extensive. We have very much appreciated the willingness of people to contribute to the text and comment on drafts in support of our ambition to produce a truly transdisciplinary account. Sandra Mather and Suzanne Yee from the University of Liverpool's School of Environmental Sciences need to be credited for their excellent graphics input. Finally, we offer our thanks to Tim Hardwick, our commissioning editor at Earthscan, for his continuing support and guidance during the preparation of the book and his willingness to vary deadlines!

Sue Kidd, Andy Plater and Chris Frid
Liverpool, July 2010

References

Agardy, T. (2010) *Ocean Zoning: Making Marine Management More Effective*, Earthscan, London and Washington, DC

McLeod, J and Leslie, H. (2009) *Ecosystem-based Management for the Oceans,* Island, Washington, DC

Potts, J. and Smith, H. (eds) (2005) *Managing Britain's Marine and Coastal Resources,* Routledge, Abingdon

Starkey, J., Holm, P. and Barnard, M. (eds) (2007) *Oceans Past*, Earthscan, London and Washington, DC

Chapter 1

The Ecosystem Approach and Planning and Management of the Marine Environment

Sue Kidd, Ed Maltby, Leonie Robinson, Adam Barker and Chris Lumb

This chapter aims to:

- Explain the origins, definitions and principles of the ecosystem approach (EA) and associated United Nations (UN) operational guidance;
- Review some of the lessons that can be drawn from existing experience of applying EA in non-marine and marine areas; and
- Provide a transdisciplinary discussion of key issues that require further attention in developing EA in planning and management of the sea in order to set the scene for the chapters that follow.

Introduction

We are in the midst of a paradigm shift in planning and management of the natural environment and the resources that are derived from the functioning of component ecosystems. This shift is based on a number of key premises which recognize that:

- sustainability of economic systems and the quality of human life is inevitably dependent on the maintenance of healthy ecosystems;
- humans are an integral part of ecosystems rather than separate from them; and
- a sectoral approach to planning and management is generally insufficient to deal with the complex interrelationships and diverse stakeholder priorities that exist in the real world.

Adoption of EA as a methodological framework for a more holistic style of planning and management that reflects these premises has been promoted in particular, but not exclusively, by the 1992 UN Convention on Biological Diversity (CBD) and has been incorporated into a growing number of other international conventions and policy documents such as the 2002 Plan of Implementation of the World Summit on Sustainable Development.

It is therefore not surprising that EA is now regarded as a central concept shaping the development of new planning and management arrangements for the sea. In terms of the international legal framework related to the marine environment, the Convention on the Conservation of Antarctic Marine Living Resources, which came into force in 1982, played a pioneering role in the development of the Ecosystem Approach (Constable et al, 2000). Since that time, EA has featured in many successor marine conventions such as the 1992 Convention on Protection of the Marine Environment of the Baltic Sea Area (Backer and Leppanen, 2008), the 1992 Convention on the Protection of the Marine Environment in the North East Atlantic, and the 1995 amendment to the Convention on the Protection of the Marine Environment of the Coastal Region of the Mediterranean. It is also increasingly advocated as an organizing concept in a wide range of marine-related policy documents. For example, in 2006 the UN General Assembly adopted Resolution 61/222 on 'Oceans and the Law of the Sea' which connected existing international instruments, in particular the UN Convention on the Law of the Sea (UNCLOS), to EA. It also endorsed the call of the 2002 World Summit on Sustainable Development Plan of Implementation to apply the ecosystem approach by 2010 in order to urgently restore fish stocks by 2015, establish representative networks of Marine Protected Areas by 2012 and achieve a significant reduction in the rate of loss of biological diversity by 2010 (Maes, 2008). This guidance has been a key factor prompting an unprecedented level of activity in marine planning and management across the globe and, in many instances, the ecosystem approach is being given prominence in associated policy documents. For example, the European Commission has identified EA as a key principle guiding the development of a new integrated Maritime Policy for the European Union (EU) (CEC, 2007). Similarly, the Interim Framework for Effective Coastal and Marine Spatial Planning in the United States, published in December 2009, emphasizes that such activity should fully reflect EA understanding (Interagency Ocean Policy Task Force, 2009). EA is also cited as underpinning the high level marine objectives issued by the UK government in 2009 as a precursor to the production of the first national Marine Policy Statement (Defra, 2009). Similar commitments to EA are made in ocean policy documents published by Canada, Columbia, Norway, Portugal and others (Intergovernmental Oceanographic Commission, 2007).

However, despite these developments, there is still considerable debate about what is actually meant by EA and how it might be applied to marine planning and management practice. This uncertainty was highlighted at the seventh meeting of the Open-ended Informal Consultative Process on Oceans and the Law of the Sea in 2006, which focused on EA and its application. Key conclusions of the meeting included the need to: demystify the concept and develop a clearer understanding of its implications; encourage more active implementation of EA in marine planning and management practice; and improve understanding and application of EA by sharing experiences and lessons learned (International Institute for Sustainable Development, 2006). This book responds to this agenda drawing on a combination of literature review, the authors' own research and practice experience and transdisciplinary discussions at a series of research council-funded seminars held between 2007 and 2009 in the UK. This chapter sets the scene for following contributions by: explaining the origins, definitions and principles of EA and associated UN operational guidance; reviewing some of the lessons that can be drawn from existing experience of applying EA in non-marine and marine areas; and providing a transdisciplinary discussion of key issues that require further attention in developing EA in future planning and management of the sea which will be explored further in subsequent chapters.

Origins of EA

The current focus on EA to environmental planning and management reflects not only contemporary understanding of environmental processes and the environmental challenges that need to be addressed, but also recognition of the shortcomings of current institutional frameworks and practices in dealing with these. For example, since Alfred George Tansley first coined the term in 1935 (cited in Wang, 2004), ecosystems have come to be a central focus in the study of ecology and environmental management. However, as understanding of the functioning of ecosystems has developed, so too has appreciation of the close interaction between humans and the environment. This has been reflected in successive UN's Earth Summits that, since 1972, have drawn international attention to the ways in which rapid population growth, increased economic activity and improved standards of living are resulting in unprecedented levels of demand for natural resources and are causing significant environmental, as well as related social and economic stress in many parts of the world. At the Earth Summits particular attention has been drawn to the impact of human pressures on the marine environment (UNEP, 2002). As awareness of these critical interconnections has increased, the inadequacies of existing environmental management arrangements have become all too apparent. Fragmented administrative structures in which policy and operational responsibilities are divided between a disparate array of organizations, narrow sectoral decision-making systems with competing and contradictory objectives, a disconnection between national, regional and local level activities and between natural and administrative boundaries, are typical features of governance in countries all over the world and particularly so in relation to marine areas. This situation both aggravates environmental problems and impedes efforts to adopt more sustainable management practices. It is within this context that EA is emerging as the dominant paradigm for tackling environmental concerns.

The origins of EA can perhaps be traced back as far as the early 20th century and the work of visionary ecologists such as Patrick Geddes, who championed the place of ecological understanding as the underpinning of sound regional planning (Allen, 1976; Kidd, 2007), and Aldo Leopold and his radical thinking about system-based approaches to land management (Bengston et al, 2001). However, as Bengston et al suggest, it wasn't until the late 1960s and early 1970s that ecosystem management approaches began to be widely advocated in the US and elsewhere, and not until the 1990s that such thinking had gained widespread support among environmental policy-makers and managers. Even by this stage its core features remained loosely defined and indeed it was argued by some (for example, More, 1996) that a single definition or model was inappropriate as the concept was too complex to be codified as a result of rapidly changing scientific understanding, professional expertise and social values. What emerged instead was a number of interrelated characteristics (including environmental, social and economic dimensions) which in differing combinations, in different contexts, were generally considered to constitute ecosystem management (see Box 1.1).

Box 1.1 *Characteristics of ecosystem management*

- Maintain ecosystem health (for example, maintain and protect ecosystem integrity and functions, restore damaged ecosystems).
- Protect and restore biodiversity (protect native genes, species, populations, ecosystems).
- Ensure sustainability (for example, incorporate long time horizons, consider the needs of future generations, include both ecological and economic sustainability).
- Systems perspective (for example, a broad, holistic approach to management; manage at multiple scales and consider the connections between different scales; coordinate across administrative, political and other boundaries to define and manage ecosystems at appropriate scales).
- Human dimensions (for example, incorporate social values and accommodate human uses within ecological constraints, view humans as embedded in nature).
- Adaptive management, in which management is conducted as a continuous experiment.
- Collaboration, in which planning and management are joint decision-making processes that involve sharing power with key stakeholders.

Source: Based on Bengston et al (2001, p473)

Table 1.1 Characteristics of traditional and ecosystem management approaches

Characteristic	Traditional approach	EA	Benefits of EA
Management structure	Isolationist	Horizontal/inclusive	More holistic (addresses multiple problems)
Management objectives	Single issue focus	Ecosystem focus	Reduces chance of cumulative effects and opposing objectives
Overarching objective	Economic/environmental trade-offs	Maintaining ecological integrity	More science-focused decisions
Management boundaries	Constitutionally defined	Ecologically defined	Reduces overlap between multiple jurisdictions
Management approach	One-size fits all	Place-specific	Objectives are relevant to particular system
Citizen engagement	Limited consultations	Extensive collaboration	Decisions are more transparent to local stakeholders and more likely to receive lasting support
Decision-making process	Linear, top-down	Integrative (both top-down and bottom-up) and circular	Better integration of multiple values increasing the likelihood of consensus
Follow-up	Limited	Adaptive management	Increased opportunity to learn from experiences

Source: Lamont (2006, p9)

Lamont (2006) provides a helpful analysis of how an ecosystem management approach differs from traditional management practices and here again highlights the importance of the human dimension (see Table 1.1).

However, following the ratification of the Convention on Biological Diversity at the UN's Earth Summit in Rio de Janeiro in 1992, there has been increasing pressure to develop a more formal definition of EA and its main features. This process has been facilitated by a series of meetings of international experts who comprise the Subsidiary Body on Science, Technology and Technical Advice (SBSTTA) to the CBD and by associated decisions by the Conference of the Parties (COP) of the CBD. Together these activities have gradually elaborated the detail of EA and more recently urged more active implementation at national and local levels. Key steps in this process are illustrated in Table 1.2.

Table 1.2 Key steps in the development of EA under the CBD

SBSTTA1 (Paris, 1995)	Recommendation 1/3: Recommends that a holistic approach be taken towards conservation and sustainable use of biological diversity and EA should be the primary framework for action taken.
COP2 (Jakarta, 1995)	Decision II/8: Reaffirms that EA should be taken as the primary framework for action.
SBSTTA2 (Montreal, 1996)	Recommendation II/1: Advocated the development of EA guidelines and indicators and identified certain priority tasks.
COP3 (Buenos Aires, 1996)	Decision III/10: Endorsed II/1 above and outlined work in thematic areas and indicators.
COP4 (Bratislava, 1998)	Decision IV/1B: Requested SBSTTA to develop principles and other guidance on EA.
COP5 (Nairobi, 2000)	Decision V/6: Endorsed the description of EA and operational guidance and recommended the application of the principles and other guidance on EA.
COP6 (The Hague, 2002)	Decision VI/12: Urged the application of EA at national and regional levels.
COP7 (Kuala Lumpa, 2004)	Decision VII/11: Agreed that the priority at this time should be on facilitating implementation of the ecosystem approach and welcomed additional guidelines to this effect.
COP9 (Bonn, 2008)	Decision IX/7: Urged parties to strengthen and promote the use of the ecosystem approach more widely and effectively and further promote collaboration and experience exchange and capacity building.

Source: Derived from Secretariat of the CBD (2010a)

Definition of EA

An important dimension of this activity has been the formalization of internationally recognized definitions of EA. Most frequently quoted, and perhaps carrying most legal weight in international law, is the definition put forward by Decision 2000 V/6 of the COP to the CBD. This defines EA as:

> *A strategy for the integrated management of land, water and living resources which promotes conservation and sustainable use in an equitable way. (CBD COP, 2000, V/6)*

The idea is that integrated management practices that follow EA should be based upon:

The application of appropriate scientific methodologies focussed upon levels of biological organization, which encompass the essential structure, processes, functions and interactions among organism and their environment ... recogniz(ing) that humans, with their cultural diversity, are an integral component of many ecosystems. (CBD COP, 2000, V/6).

It is envisaged by the COP that the EA can act as a framework for balancing and integrating the three objectives of the CBD which relate to:

- conservation of biological diversity;
- the sustainable use of its components; and
- fair and equitable sharing of the benefits arising out of the utilization of genetic resources (Glowka et al, 1994).

EA recognizes the need to tackle conservation and environmental issues from an integrated perspective which combines closely interconnected economic and social considerations. Figure 1.1 provides an illustration of what is envisaged.

Figure 1.1 Conceptualization of EA

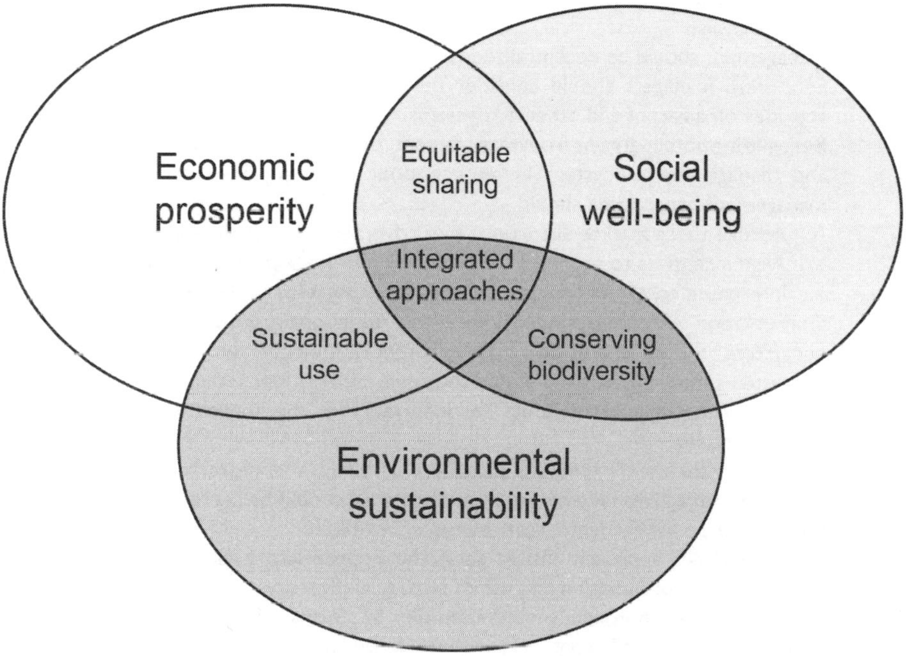

Source: Maltby and Crofts in Maltby (2006)

EA principles

Underpinning this broad CBD definition, a series of 12 complementary and interlinked EA principles have been identified (see Box 1.2) which highlight the complexity of the concept. Following a workshop of international experts in Lilongwe in 1998, a draft set of principles (known as the Malawi principles) were developed that drew on the outcome of the Sibthorp Seminar on Ecosystem Management (Maltby et al, 1999). These were refined by SBSTTA to the CBD at its meeting in Montreal in 2000, and together with associated guidance, they were finally adopted by the COP to the CBD at its fifth meeting in May 2000 (decision V/6).

As can be seen from Box 1.2, the 12 principles could be considered to be arranged in a somewhat arbitrary order and the International Union for Conservation of Nature's (IUCN) Commission on Ecosystem Management (CEM) has tried to facilitate understanding by grouping them into five steps which are arranged in a roughly chronological sequence (see Table 1.3). This development is helpful in drawing attention to the key areas for consideration and action.

Box 1.2 *CBD: EA principles*

1. The objectives of management of land, water and living resources are a matter of societal choice.
2. Management should be decentralized to the lowest appropriate level.
3. Ecosystem managers should consider the effects (actual or potential) of their activities on adjacent and other ecosystems.
4. Recognizing potential gains from management, there is usually a need to understand and manage the ecosystem in an economic context. Any such ecosystem-management programme should:
    –   Reduce those market distortions that adversely affect biological diversity;
    –   Align incentives to promote biodiversity conservation and sustainable use;
    –   Internalize costs and benefits in the given ecosystem to the extent feasible.
5. Conservation of ecosystem structure and functioning, in order to maintain ecosystem services, should be a priority target of the ecosystem approach.
6. Ecosystems must be managed within the limits of their functioning.
7. The ecosystem approach should be undertaken at the appropriate spatial and temporal scales.
8. Recognizing the varying temporal scales and lag-effects that characterize ecosystem processes, objectives for ecosystem management should be set for the long term.
9. Management must recognize that change is inevitable.
10. The ecosystem approach should seek the appropriate balance between, and integration of, conservation and use of biological diversity.
11. The ecosystem approach should consider all forms of relevant information, including scientific and indigenous and local knowledge, innovations and practices.
12. The ecosystem approach should involve all relevant sectors of society and scientific disciplines.

Source: CBD COP (2000, V/6)

Table 1.3 The 12 EA principles, grouped into five steps by IUCN's CEM

Step A. Key stakeholders and area		
Stakeholders	Principle 1	The objectives of management of land, water and living resources are a matter of societal choice.
	Principle 12	The EA should involve all relevant sectors of society and scientific disciplines.
Area analysis	Principle 7	The EA should be undertaken at the appropriate spatial scale.
	Principle 11	The EA should consider all forms of relevant information.
	Principle 12	The EA should involve all relevant sectors of society and scientific disciplines.
Step B. Ecosystem structure, function and management		
Ecosystem structure and function	Principle 5	Conservation of ecosystem structure and function, to maintain ecosystem services, should be a priority.
	Principle 6	Ecosystems must be managed within the limits of their functioning.
	Principle 10	The EA should seek the appropriate balance between, and integration of, conservation and use of biological diversity.
Ecosystem management	Principle 2	Management should be decentralized to the lowest appropriate level.
Step C. Economic issues		
Principle 4		There is usually a need to understand and manage the ecosystem in an economic context and to: i) reduce market distortions that adversely affect biological diversity; ii) align incentives to promote biodiversity conservation and sustainable use; and iii) internalize costs and benefits in the given ecosystem.
Step D. Adaptive management over space		
Principle 3		Ecosystem managers should consider the effects of their activities on adjacent and other ecosystems.
Principle 7		The EA should be undertaken at the appropriate spatial scale
Step E. Adaptive management over time		
Principle 7		The EA should be undertaken at the appropriate temporal scale.
Principle 8		Recognizing the varying temporal scales and lag-effects that characterize ecosystem processes, objectives for ecosystem management should be set for the long term.
Principle 9		Management must recognize that change is inevitable.

Source: Shepherd (2008)

Operational guidance for EA

In addition to the 12 EA principles, the COP has also put forward five points of operational guidance. These are explored below.

1. Focus on the functional relationships and processes within ecosystems

Many factors determine the health of ecosystems and their ability to withstand stress. Improved knowledge of the functions and structure of ecosystems is needed to understand (i) ecosystem resilience and the effects of biodiversity loss (species and genetic levels) and habitat fragmentation; (ii) underlying causes of biodiversity loss; and (iii) determinants of local biological diversity in management decisions. However, it is recognized that ecosystem management has to be undertaken in the absence of perfect knowledge, and dialogue between different parties is needed to come to a considered view as to the appropriate way forward.

2. Enhance benefit-sharing

Ecosystems provide the basis of human environmental security and sustainability and EA seeks to ensure that the benefits derived from the functioning of ecosystems are maintained or restored. However, in addition to the scientific challenges presented by such a task, there are also important social challenges to be addressed here. Limited public understanding of environment/human interactions, and the fact that ecosystem goods and services are, in many instances, external to the market economy or lack proper market valuation, is thought to hamper effective planning and management of ecosystems. Central to EA, therefore, is a need to develop a deeper awareness of human/environment connections and benefits and to reduce or remove incentives for environmental degradation.

3. Use adaptive management practices

Given the complexity and variability of ecosystem processes, the constant flux of environment/human interactions and imperfect scientific understanding, it is important to encourage a flexible, adaptive management or 'learning by doing' attitude in which experimentation, monitoring and adjustment are key features.

4. Carry out management actions at the scale appropriate for the issue being addressed, with decentralization to lowest level, as appropriate

Ecosystems function at a variety of scales depending upon the issues being considered and management structures and actions should reflect this situation. The need to adopt appropriate management practices requires that effective responsibility should lie at the local level. However, it is important that local actions are supported by appropriate policy and legislative frameworks at regional, national and international levels and should encompass the practices of all relevant stakeholders.

5. Ensure intersectoral cooperation

The complexity of ecosystem functioning requires a partnership approach to planning and management. This needs to encourage cross-sectoral coordination between different aspects of public policy, for example nature conservation, agriculture, forestry and fisheries and indeed other public policy areas such as land-use planning and economic development. It also should facilitate inter-agency cooperation between government departments and agencies and private and voluntary sector interests.

Experience of applying EA in non-marine contexts

Having briefly charted the development of EA and summarized key features of the concept, consideration is now given to how these ideas have been translated into practice and to distilling some lessons from experience to date. An initial overview can be obtained from the findings of an in-depth review of the application of EA which was undertaken by the SBSTTA between 2004 and 2007 and reported to the ninth meeting of the COP in 2008 (SBSTTA, 2007). This concluded that, although there was evidence that aspects of the ecosystem approach were being adopted by many countries, relatively few had practical experience of direct application of most of the principles. Equally, the review indicated that it was difficult to ascertain the extent to which EA-related activity might favour particular biomes, but drawing upon an analysis of the CBD Ecosystem Approach Source Book (Secretariat of the CBD, 2010b) it concluded that some biomes were particularly poorly represented by case studies, for example islands, mountains and polar regions. These findings informed one of the report's most stark conclusions, which is that that the ecosystem approach is not as yet being applied systematically enough to reduce the rate of global biodiversity loss.

Looking in the first instance at experience of applying EA in non-marine settings, it appears that so far a prominent area of application of the approach has been in relation to water management. This has been encouraged, for example, by international policy papers associated with the Ramsar Convention (The Swiss Agency for the Environment, Forests and Landscape et al, 2002) which is reflected in activities in many countries including Australia (Hillman et al, 2003) the US (Hartig et al, 1998; Klug, 2002) Asia, Africa and South America

(Smith and Maltby, 2003). Forest areas have also been a focus of attention, for example, in Indonesian Papua and the Northern Congo (Shepherd, 2008). More wide-ranging application of the approach is evident in countries such as the US (Bengston et al, 2001) particularly through the work of the US Fish and Wildlife Service (Danter et al, 2000) and the UK (Laffoley et al, 2004).

Insights gained from these experiences suggest that even in the US, which perhaps has the most longstanding experience of the approach, understanding of what is meant by EA is still not widespread. However, there does appear to be a generally sympathetic public attitude towards its ambitions here and elsewhere, and it clearly has growing professional support. The focus of attention has therefore quickly moved from discussion about 'what is an ecosystem approach to planning and management and should it be implemented?' to 'how can it be put into practice?' (Bengston et al, 2001). Interestingly the same authors highlight the dynamic nature of the concept with increasing recognition being given to the place of social values and public involvement (in addition to scientific understanding) in translating EA into planning and management activities in the ground. This emphasis on application and the need to tailor and interpret the broad normative framework provided by EA principles into specific environmental, social and economic contexts is also a key point made by the SBSTTA in its practice review report. This identifies wider application of EA and a focus upon learning by doing as a key priority at the present time (SBSTTA, 2007).

The importance of institutional design to support EA also emerges as a significant theme. For example, Hillman et al (2003), writing about experience in the Lachlan catchment in Australia, stress the role of institutional structures and processes in nurturing 'social capital', developing trust, inclusiveness and mutual respect between stakeholders and the importance of these features to the success of EA initiatives. They also argue that the time and effort involved in building partnerships and achieving consensus in decision-making is rewarded by stronger commitment among stakeholders to the planning proposals and management measures adopted. Fee et al (2009) reiterate these issues with reference to Canadian and German experiences, but also draw particular attention to the role of politicians in taking forward EA in an effective way. They indicate that resource planning and management decision-making can become stuck in the political arena and consider that institutional structures that combine both top-down and bottom-up inputs are key to effective delivery of EA. It is interesting to note that, in the summary of the general barriers to the use of EA identified by the SBSTTA in their practice review (see Box 1.3), nearly all barriers highlighted have an institutional design dimension.

The challenges of adaptive management are another recurring theme with both technical and stakeholder management issues prominent. For example

---

Box 1.3 *General barriers to the use of EA*

- Ineffective stakeholder participation in planning and management
- Limited understanding of what the approach seeks to achieve
- The lack of capacity for decentralized and integrated management
- Insufficient institutional cooperation and capacity
- The lack of dedicated organizations able to support delivery of EA
- The overriding influence of perverse incentives
- Conflicting political priorities, including those that arise when a more holistic approach to planning is adopted

---

Source: SBSTTA (2007, para 27)

Hartig et al (1998) warn that sustained stakeholder involvement and support are dependent on the ability to demonstrate measurable progress toward clear objectives and targets. They suggest that emphasis must be placed on a step-wise approach to goal attainment in which both short- and long-term milestones need to be set out. They also recommend that progress should not only be achieved, but also documented and celebrated if multiple stakeholders are to be kept on board. However, this emphasis on clear targets and goal attainment (in order to keep stakeholders engaged) does present challenges in situations where data and understanding may be limited. Management styles that recognize uncertainty (notably partial scientific understanding) and support action learning are considered to be critical to the application of EA, but also a source of difficulty for those looking for clearly defined outcomes, predictability and measures of success to garner community support (Hartig et al, 1998; Hillman et al, 2003). The same authors indicate that the adaptive styles of management required by EA can present significant challenges to organizations which often need to shift from an emphasis on control to an emphasis on responsiveness. They suggest that, in the spirit of adaptive management, change must be seen as a continuous process. Stable, predictive and predictable organizational processes need to be replaced by adhocracy, and leadership becomes a critical feature of success in these circumstances (Danter et al, 2000). Here it is suggested that EA requires management processes that focus on cooperative learning, harness the instinctive learning of front line staff, view management actions as experiments, provide opportunities to practice and make errors, and see assessment, monitoring and research as essential elements. Danter et al (2000) warn that this continuous organizational renewal is time and energy intensive for leaders and that this needs to be acknowledged in devising EA institutional structures and management arrangements. Finally, Klug (2002) makes some interesting observations about the gap between traditional environmental laws and policies which tend to be based on the past equilibrium paradigm of ecosystems and current 'non-

---

Box 1.4 *Characteristics of successful implementation of EA*

- The development of a management plan
- Good stakeholder involvement
- Good public awareness
- Good cooperation among stakeholders and agencies
- Good communication amongst stakeholders
- Good information sharing
- Adequate personnel resources
- Adequate funding
- The availability of scientific information
- Subsequent changes in the management of activities

Source: Turner, 2004 cited in Laffoley et al (2004)

equilibrium' understanding and acceptance of change which is implicit in EA. He suggests that legal reform may be necessary to reflect this new view.

These themes and others are evident in the listing (shown in Box 1.4) of key characteristics of successful implementation of EA that emerged from a review of early UK experience. Interestingly, there is again a strong focus upon institutional and process design factors rather than scientific characteristics here.

Experience of applying EA in the marine environment

As mentioned at the beginning of this chapter, the adoption of EA in environmental planning and management has by no means been confined to terrestrial areas. Indeed there has been particularly enthusiastic experimentation with the concept in the marine environment. By 2004 for example, 16 Large Marine Ecosystem (LME) management projects were in existence in Asia, Africa, South America and Europe involving a total of 126 countries, many of which sought to incorporate EA thinking in their activities (Wang, 2004). These projects have been financially and technically supported by the Global Environment Facility, the World Bank, the UN Industrial Development Organization, the Intergovernmental Oceanographic Commission, the World Conservation Union, and others. Today, more than 140 countries are participating in the United Nations Environment Programme's (UNEP) Regional Seas initiative covering the Black Sea, Wider Caribbean, East Asian Seas, Eastern Africa, South Asian Seas, Regional Organization for the Protection of the Marine Environment (ROPME) Sea Area, Mediterranean, North-East Pacific, North-West Pacific, Red Sea and Gulf of Aden, South-East Pacific, Pacific, and Western Africa (UNEP, 2010).

Box 1.5 *Necessary functions for management of LMEs*

- Determination of boundaries of the relevant ecosystem.
- Assessment of resources in the ecosystem and the development of an understanding of the ecological balances encompassed therein.
- Appraisal of the varying human uses of the area of the LME and the relevant interplay with one another and with the environment.
- Establishment of goals, objectives and priorities for resources and the environment of the LME, taking systematic account of scientific data and socio-economic considerations.
- Regulation of activities affecting the LME so that activities conform to choices and priorities.
- Shaping suitable institutional machinery and governance arrangements for policy-making and administration of the LME uses.
- Oversight, evaluation, monitoring and assessment of activities in an effective manner so as to allow for needed changes in management efforts and objectives.

Source: Developed from Juda (1999)

Reflecting this growing body of experience, a number of authors have sought to distil some of the key lessons that have emerged so far in applying EA to planning and management of the marine environment. For example, as far back as 1999, Juda set out what he considered to be important considerations in applying ecosystem-based management principles to the governance of LMEs. A summary of his analysis is set out in Box 1.5.

The simplicity of Juda's analysis is both helpful in identifying key areas of 'process' consideration, but also of rather limited value in unravelling the complexity of issues inherent in each 'step' along the way. In particular, in contrast to much of the discussion set out in the previous section, the human/institutional dimension seems somewhat underplayed. A richer and complementary view of the challenges posed in operationalizing EA in the context of the marine environment is, however, presented in a 2004 report published by English Nature (Laffoley et al, 2004). This identifies different areas where coherence in action is needed. Figure 1.2 illustrates how these different areas of coherence relate to one another, while Table 1.4 sets out priorities for action under each heading.

In addition to such efforts to further develop conceptual understanding of EA and its relation to planning and management of the marine environment, a growing body of literature has also begun to articulate some of the challenges involved in its application.

For example, Frid et al (2006) have identified various barriers to delivering ecosystem-based fisheries management (Box 1.6) most notably the deficiency in our understanding of the complexity and dynamics of marine ecosystems, including human aspects of it.

Table 1.4 EA in marine and coastal environments – seven areas of coherence and priorities for action

---

Environmental Coherence

Taking a fully representative approach to biodiversity
Using surrogate information sources
Defining ecosystem outcomes being sought
Avoiding damaging the genetics of species
Implementing strict site protection measures

Economic Coherence

Defining economic objectives
Developing management effectiveness indicators
Using best practice for assessing environmental impacts
Addressing combined and cumulative impacts
Fishing within ecosystem limits
Taking an integrated approach to nutrient enrichment

Social Coherence

Stakeholder participation and transparency in decisions
Planning decision-making processes
Effective participation by all stakeholders
Understanding and ownership of biodiversity benefits

Spatial Coherence

European Marine Strategy spatial framework
Implementing a spatial planning framework
Spatial regulation and management of the resource
Spatial distribution of the resource
Providing a common baseline and bathymetry data set

Temporal Coherence

Working with 'locked-in' changes to the environment
Working with past impacts and 'shifting baselines'
Sustaining long-term political ambition
Establishing a timeframe-relevant indicator set
Implementing a regional sea management timetable

Scientific Coherence

Aligning science to society and sustainable development
Undertaking regional sea-scale science
Improving access to data
Widening the scope of scientific advice
Supporting greater ownership and use of advice
Improve the synthesis of existing science

Institutional Coherence

Reforming institutional arrangements
Providing high-level support and coordination
Providing adequate support at local levels

Source: Laffoley et al (2004)

Figure 1.2 EA in marine and coastal environments – areas of coherence

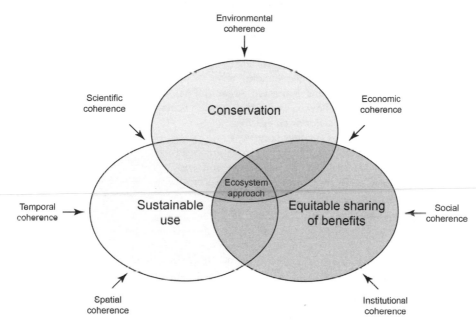

Source: Laffoley et al (2004)

Frid et al's analysis is also reflected in the conclusions of a number of other authors. For example, Wang (2004), in his commentary on LME management experience, notes that scientific understanding of the functioning of marine ecosystems is still limited in many areas and that in his view there is an insufficiently strong scientific basis to support EA to LMEs. High levels of scientific uncertainty also feature as a source of concern in Hewitt and Low's (2000) account of fisheries management in the Antarctic, as does the high cost of new research to address areas of limited understanding. Given inevitable resource constraints, they fear that scientific uncertainty may be used as a reason for opposing management measures (such as catch restrictions) and that short-term economic interests may lead to a decoupling of scientific understanding and policy/decision-making. The consensus decision-making aspects of EA they suggest may lead to a 'lowest common denominator' approach to management with 'political'/'economic' agendas dominating. Overall they argue for a greater emphasis on the precautionary principle within EA.

A more encouraging picture of how to make progress despite limitations in scientific understanding is, however, presented by Day (2008) in his discussion of the role of monitoring in the adaptive planning and management of Australia's Great Barrier Reef. Day sets out a number of key lessons including recognizing the value in: starting with a modest monitoring programme; being prepared to

---

Box 1.6 *Barriers to delivering EA-informed fisheries management*

Lack of critical understanding related to:

- links between hydrographic regimes and fish stock dynamics;
- the importance of habitat distributions;
- 'design rules' for Marine Protected Areas (MPAs), how they should vary by regions/systems, and how to ensure that they are ecologically linked;
- ecological dependence/foodweb dynamics;
- predictive capabilities in complex systems;
- how to incorporate uncertainty into management advice;
- the genetics of target and non-target organisms; and
- the response of fishers to management measures.

Source: Frid et al (2006)

combine a mix of monitoring methods and be innovative; determining who is best able and suited to undertake monitoring; and, wherever possible, encouraging stakeholder participation in the monitoring and review process. Importantly, he stresses that it is not practical to wait for perfect science before taking planning and management action.

Wang (2004) and Hyun (2005), writing about management experience in the Gulf of California, raise the issue of ecologically linked regions and systems and the major jurisdictional difficulties posed by adopting EA in marine management arrangements. The 'natural' boundaries of ocean ecosystems will cross over state, provincial, regional and municipal borders, existing politically agreed maritime zones and established divisions between land/sea jurisdictions. Effective LME management therefore not only requires coordination of marine and landward activities but also global or macro-regional seas mechanisms. Wang considers that the logical extension of EA would therefore result in institutional requirements that would be too expensive, too clumsy and too politically divisive to be operationally effective. Instead he suggests that optimal LME management arrangements should reflect a synthesis of geographic, scientific and political elements or what he describes as the geopolitical ecosystem (Wang, 2004). On a related theme, Wang also notes that the ecosystem approach to LME management may not only be unable to resolve existing ocean management problems but may also magnify their complexity. In establishing new management arrangements it may simply add to the mushrooming of institutions and plans with overlapping and competing functions, duplicate effort and add to waste and inefficiencies.

Again, however, alternative more optimistic perspectives can also be found. Within the European context for example, the Helsinki Commission

(HELCOM) Baltic Sea Action Plan is a model that many are looking to for inspiration. Backer et al (2009) in a reflective account of the application of EA in the Baltic discusses how the plan is a contract between coastal country governments and the European Commission, to commit themselves to carry out specific actions for achieving agreed ecological objectives, and eventually Good Environmental Status for the Baltic Sea by 2021. The plan is considered to have been particularly successful in establishing a framework for integrated action on a range of pollution concerns. On a less positive note, as yet individual country pollution reduction targets remain provisional and the ability of the plan to deal effectively with a range of important agricultural and fisheries matters are limited by wider EU responsibilities in these spheres. However, even though it is evident that the plan is not a final recipe for a clean Baltic Sea, the authors conclude that it is an important step in an adaptive management process that has been ongoing since the signing of the Helsinki Convention in 1974. As with any new public policy concept, EA has begun to evolve through debates and battles related to definitions, fundamental principles and policy implications. Backer et al suggest, 'regardless of the outcome of such battles in terms of final substance given to wordings like Ecosystem Approach, or Good Environmental Status, a positive aspect of such new concepts lies in the momentum they carry' (Backer et al, 2009, p649).

Key issues in the development of EA in marine planning and management

What is evident from the above discussions is that EA is the prevailing framework now guiding the development of the new approaches to marine planning and management. These are clearly required if we are to support healthy and productive seas in a context where human pressures on the marine environment are growing rapidly. However, it is also evident that application of EA poses significant challenges. This chapter concludes with a transdisciplinary discussion of some of the key issues that require further attention in developing EA in future planning and management of the sea. These issues are grouped under the following themes which will be developed more fully in subsequent chapters:

- Developing the human dimension.
- Addressing key information challenges.
- Connecting to wider agendas.

Developing the human dimension

Stressing the holistic ambitions of EA

Arguably one of the key challenges in operationalizing EA relates to the term itself. It is evident from the literature that (like sustainable development) many different interpretations of EA exist, despite CBD-related attempts to clarify the position. This is not helped by the rather loose and overlapping use of terminology in many instances. For example, EA, ecosystem management, ecosystem-based management and other similar terms are often considered to be interchangeable with EA, but may have narrower, less holistic meanings in particular contexts. As a result, it is not surprising that some have come to regard the main focus of EA as the better alignment of planning and management units to the boundaries of natural systems and the application of rigorous scientific understanding of ecosystem dynamics to the development of associated planning and management approaches. While such a view is not inconsistent with EA, it is only partial and does not reflect the much more strongly human-orientated view which is implicit in CBD interpretations.

Examination of the EA definition, principles and operational guidance issued by the CBD COP reveals not only that humans should be regarded as an integral component of ecosystems, but also that, at its heart, EA is a framework for the design of management structures and processes that encourage understanding of human/environment relationships and help guide them in a sustainable manner. Given this, it is perhaps unsurprising that there have been some difficulties in engaging fully with the more traditionally oriented scientific community and gaining their acceptance of this broader interpretation of EA. Similarly, for the non-scientific community who may know little of the CBD, EA may seem indistinguishable from established environmentally focused planning and management practices and therefore appear of little relevance to them. 'Ecosystem' inevitably suggests a Tansley-type interpretation and approach. The much wider interpretation, captured by the CBD, seems to many scientists either irrational or at best difficult to understand and is often hidden altogether from the wider, non-scientific community. While it might be useful to consider a modification of the term such as 'ecosystem approach to management', resistance has been based on the further confusion that this might cause. Continuing dialogue and education about the holistic nature of EA among marine-related stakeholder communities therefore seems to be critical.

Human activity as the focus of planning and management action

Following on from this, it is apparent that emphasizing the human dimension of EA is particularly important as this is not only a critical feature of the concept, but also one that is arguably least understood or accepted. Looking deeper into this, again the terminology proves unhelpful, as one of the problems associated with the translation of EA, is where it is interpreted literally as meaning an approach where we try to manage ecosystems. If we are to move forward, it is important to recognize that we cannot manage ecosystems (particularly marine ecosystems); we manage the human activities that cause pressures in them, not the constituent components of them. For example, it is incontrovertible that fishing has caused changes in particular marine ecosystems. However, where this has occurred, we cannot even attempt to manage the complexities of the ecosystem itself, we must use our understanding of how fishing activity has driven change in the system, to try to reduce or reverse impacts through management of the activity (for example, reduction of fishing effort or implementation of closed areas or seasons). Clearly, in doing so, we must also understand the natural variability of ecosystems, the processes driving this variability, and the linkages between the components, key tenets of implementing an EA as defined in CONSSO (2002). Without doing so, we have little chance of interpreting the likely response of ecosystem components to changing human pressures. Nevertheless, this is different from suggesting that we can manage the variability in the system itself. Accepting this difference may help to reduce the resistance to EA by some in the scientific community.

The cutting edge of EA-informed marine planning and management should therefore be clearly identified as the direction of human activity in a manner that is consistent with the health of our seas. In line with this, it seems to be critical that detailed information about the existing pattern of human use of the sea needs to be developed in parallel with environmental data in order to assess where change in human activity may be beneficial and future patterns of human activity might be most benign. A balanced approach to data gathering might include developing an understanding of both direct (for example, fishing, energy generation, sea transport) and indirect (for example, sink for wastes, tourism setting, climate change amelioration) sea use and the associated economic and social value derived from this by landward communities. As Principle 11 suggests this may entail drawing upon a wide range of data sources *including indigenous and local knowledge*. A future view is also critical for marine planning in particular, as the combined pressures of global population growth, increasing human demands for natural resources, potential resource scarcity and climate change are all likely to result in a rapidly changing and increasing pattern of human dependency on the sea over the course of the next century. Anticipating and guiding these human pressures by forward-looking marine spatial planning, which includes developing and assessing future projections and alternative scenarios of human use and environmental change, will also be important and is the subject of discussion in Chapter 2.

In enhancing understanding of the human component of marine ecosystems it is worth revisiting the work of authors such as Endter-Wada et al (1998), Scoones (1999) and Marten (2001) who have put forward ideas about how the human component of any given ecosystem might be investigated. Endter-Wada et al (1998), for example, propose a framework for analysis, based on US experimentation, which considers both the nature of social science data and the role of different assessment techniques in ecosystem planning and management. Central to the framework is recognition of the need to not only determine the structural attributes of society and economy, but to also develop a working appreciation of trends, attitudes, beliefs and values. These wider attitudinal and behavioural attributes are seen as not only essential to appreciating the way in which humans 'condition' and are 'conditioned by' natural systems but, as Scoones (1999) and Holling (2001) also argue, are vital for the development of integrated management solutions that are implementable.

There is, however, evidence that the need to develop a more rounded understanding of marine-related systems encompassing the human dimension is in fact now quite widely recognized. The EU Integrated Coastal Management Demonstration Project (CEC, 1996), for instance, identified the main socio-economic dynamics impacting upon the EU coastline, whilst a Marine Spatial Planning Pilot project (MSPP Consortium, 2006) attempted to identify resource-use interactions across the North Sea. It is arguable, though, that although such research initiatives have raised the profile of social structures and behaviour in the marine environment they serve to highlight the presence of social systems, rather than the relationship between social and natural systems. It is this integrative perspective that would appear to be the biggest challenge to effective EA application to the sea.

Developing objectives that reflect societal choice

Nowhere is this challenge greater than in the definition of planning and management objectives for the marine environment. Principle 1 set out by the CBD COP stresses that 'the objectives of management of land and water and living resources are a matter of societal choice'. It is evident that without clear objectives, approaches to marine planning and management lack strategic direction and purpose, but definition of suitable objectives that can truly be said to reflect 'societal choice' is by no means an easy task and the EA, which is essentially process-oriented, offers no direct guidance here. In many countries, the separation of environmental planning and management activities from often more politically prominent socio-economic policy and plan-making, makes broad engagement in environmental planning difficult. This situation is compounded in the context of the marine environment because the effective community is somewhat detached from the area to be planned or managed, and because social and economic needs may be ill-defined and understanding of the linkages between human activities and environmental degradation of the sea may be limited. As a result, there is a danger that the objectives set may largely reflect the concerns of those with an environmental and/or a scientific perspective and predominantly focus upon environmental matters and on the marine area only. In doing so, marine plans may fail to engage effectively with the very social and economic forces (many of which will be land-based) that need to be modified if environmental objectives are to be met. As indicated above, the cutting edge of EA-informed marine planning and management is the direction of human activity. Therefore a major challenge to the delivery of the EA in the context of marine areas is achieving a rounded definition of objectives that reflect broader economic and social needs and aspirations and hopefully aligning these with those related to the sustainability and good health of the sea. An illustration of the difficulties posed is the focus of Chapter 4.

Stakeholder engagement

With these considerations in mind, effective stakeholder engagement is obviously a central aspect in the delivery of EA in marine areas and this is reflected in Principle 12 which states 'the ecosystem approach should involve all relevant sectors of society and scientific disciplines'. However, a number of barriers to effective engagement are apparent. As discussed above, problems of weak identification with marine management concerns and therefore limited interest in engagement, highlight the importance of wider social learning about marine matters and suggest that a period of informal engagement and capacity building may need to precede more formal setting of marine planning and management objectives and partnership working. Stakeholder engagement in the new EA era of integrated and holistic marine planning and management may also run up against the inertia inherent in the established sectoral pattern of plan-making and existing patterns of engagement. For example, established single sector management arrangements that give prominence to particular sectoral interests may need to be subsumed within the broader approach envisaged by EA. This is a change that may not be universally welcomed and a period of transition may be needed. In the context of the UK, for example, it seems only appropriate that negotiations about fisheries quotas and 'no take' zones should be brought within the remit of the proposed new marine plans, but there has been a strong resistance to this from the UK fishing community until parallel arrangements are in place across all EU member states. Beyond such issues there is also the matter of consultation fatigue among land-based interests who may already be asked to make an input to an increasing array of public sector policy documents, plans and initiatives. Here again, effective integration between plans may yield benefits, particularly if this extends to shared consultation arrangements and pooling of consultation responses.

Addressing key information challenges

Informing societal choice

Developing objectives to guide marine planning and management activity will entail the weighing up of different economic, social and environmental factors and inevitably (given the previous comments) rest upon judgement about which human needs, activities and ambitions will be prioritized and which will not. Objectives (and consequent planning and management measures that flow from these) will have direct and indirect, and short and longer term impacts on the pattern of human activity and well-being. This is highlighted in the reference made in Principle 1 to the need to measure the 'tangible and intangible' benefits of planning and management activity in a 'fair and equitable way' and it may be anticipated that issues of intra-generational and inter-generational equity will become increasingly to the fore in marine-related decision-making over the course of the 21st century as human pressure mounts.

A central concern of EA, however, is that stakeholders who influence the decision-making process should have the ability to make informed choices and it is important therefore that good scientific advice, developed in tandem with socio-economic understanding, is available. Data gathering on its own is insufficient, however. The real requirement is to translate this understanding into relevant planning and management responses and to produce an informed assessment of where the environmental, economic and social costs and benefits of particular management strategies lie, so that stakeholders have a clear understanding of what is involved. Again this is not an easy task. For example, understanding the benefits of reversing the decline in population size of a threatened species that is consumed by humans (for example, cod) is tangible, but considering this against the benefits of reducing the loss of a particular type of sub-tidal marine habitat to human development in order to protect the wider system integrity and in response to climate change concerns (conceivably less tangible objectives), seems unlikely to be made in a 'fair and equitable way', unless those making the decisions are well-informed of the issues and impacts associated with further degradation of either.

Improved methods of assessing and measuring (either quantitatively or qualitatively) the tangible and intangible benefits of different planning management measures on the ecosystem and the stakeholder communities, and of communicating this understanding to decision-makers are both areas where progress and imaginative new approaches are required and this is the focus of Chapters 4 and 5. As described in Principle 4, successful communication of costs and benefits is likely to involve, among other things, a translation of biological changes into an economic context and the application of ecosystem services approaches is a key area for further development here.

Spatial dynamics and different planning and management responses

Critical to both the development of effective planning and management measures and to the assessment of the distribution of costs and benefits associated with these will be an understanding of spatial dynamics. For example, it is widely accepted that ecosystem processes operate at different spatial scales and that, due to the different spatial dynamics of particular components of marine ecosystems (for example, benthic invertebrates versus marine mammals), the consequences of managing human pressures (for example, land run-off of nutrients from farming) in a particular planning unit, will also vary for the constituent components. Recognition of this is essential in guiding the formulation of marine planning and management activities that strive to follow EA (operational guidance point 4). This complexity means that it is likely that a number of different types of planning and management responses will be needed in relation to any marine area. For example, while some pressures are easily defined in terms of their spatial footprint and are therefore suitable for zoning type responses (i.e. the habitat change and smothering associated with an aggregate dredge disposal site), others are not (i.e. alien species introductions). Further research and guidance exploring the implications of spatial dynamics for the range planning and management options that may be relevant in marine areas and the circumstances in which they may be applicable would therefore be valuable. In terms of the science required to assist here, there is further work required, particularly in areas such as ecological connectivity (see Principle 7), where there is still much to be learnt in marine systems. For progress in these areas, access to good spatially resolved data for both ecological components and human activities is essential.

Temporal dynamics: The importance of a long-term view and adaptive management

In the same way, the temporal dynamics of marine ecosystems are also critical to marine planning and management activities. Understanding the limits and variability in functioning of marine ecosystems (Principle 6) requires the availability of long-term data. It is essential that as we progress into the new era of marine planning, funding the collection and analysis of long-term data sets is resourced appropriately. It is only through the interpretation of such data sets that we can advise on the limits of the ecosystems in which we are trying to manage human activities. Continued funding and use of such work will require a shift away from short-term thinking that usually tends to predominate at present particularly in relation to budgets (Principle 8).

It is also important to recognize the varying temporal scales that ecological processes operate at in considering how an adaptive management approach might work (Principle 9). The recovery or return times of different ecosystem components vary considerably (from hours to centuries) and even for specific species, the response to different pressures will vary, due to particular lag-effects associated with the mechanism through which the pressure affects it. For example, if human activity causes an acute local effect (for example, high mortality of adult fish in a particular area due to an oil spill), the response time is likely to be different to that related to an activity that causes a widespread and chronic effect (for example, reducing reproductive success of all populations in a large geographic area). If truly adaptive management measures are to be developed and implemented (operational guidance point 3), they will need to be flexible enough to account for the likely differential temporal responses observed in any given ecosystem. Again further research and guidance exploring the implications of spatial dynamics for the range planning and management options that may be relevant in marine areas and the circumstances in which they may be applicable would therefore be valuable.

Understanding structural and functional biodiversity

A key tenet of EA is that biodiversity 'encompasses the essential structure, processes, functions and interactions among organisms and their environment' including man (CBD COP, 2000, V/6). Understanding the relationships between structure and function, and the limits and behaviour of these, underlies much of the current research effort in marine science. While it is recognized that biological diversity is 'critical both for its intrinsic value and because of the key role it plays in providing the ecosystem and other services upon which we all ultimately depend' (Principle 10), science must continue to research and advise on the ecological significance of particular changes in structure and function (see operational guidance point 1). For example, how does the extirpation of particular taxa affect overall functioning of the ecosystem of which it is part? As recognized in Principle 5, simple protection of single species may be of less importance for the long-term sustainable use of biological diversity than the overall maintenance of ecosystem functioning.

It is essential that, as this is a developing area of scientific effort, the means are provided for incorporating new understanding of ecosystems' functioning and resilience, and the affects of different pressures on this, into the adaptive planning and management scenarios that are emerging at the present time.

Dealing with complexity and uncertainty

It is evident from the above discussion, that marine ecosystems are intensely complex, and in many aspects and in many areas understanding is as yet ill-developed. This is true in relation to both the human and non-human components of the system and of associated interactions. In most marine situations, significant improvements in data and in understanding of system dynamics are needed in order to reduce uncertainty and to provide a sound evidence base for decision-making. However, this raises the question about how much information do decision-makers need in order to make an effective decision? These are issues shared by all those responsible for managing complex systems and indeed such concerns are not new. For example, during the 1960s, the emergence of 'systems planning' within the field of town and country planning led to similar dilemmas. Proponents of this approach, such as McLoughlin (1969), argued for a rational comprehensive approach to planning for urban environments that entailed the careful identification, assessment and management of the complex web of interactions between people and place within the 'urban system'. The approach relied on extensive and detailed analysis geared towards the understanding of the fine grain of urban areas. The extent to which such a process could ever be achieved, however, had been challenged by Lindblom (1959) who, by the late 1950s, had already suggested that comprehensive approaches to the management of complex systems were doomed and that all we could ever do was 'muddle through' the mire of information and make decisions on an incremental basis. In the field of ecological economics, Holling (1978) similarly called for a flexible approach in dealing with complex environmental problems. In particular he advocated the use of a predictive modelling approach based on the use of information as and when it came to light. This, he later conceded, was the only way to tackle an entity that he viewed as a constantly 'moving target' (Holling, 1993). Although it must be recognized that considerable advances in data gathering and modelling have been made in recent times and are ongoing, the issue of how best to deal with complexity, uncertainty and data collection remain very live. The EA provides a strong steer here in stressing in Principle 8 that planning and management must *recognize that change is inevitable*, and in operational guidance point 3 that an adaptive or 'learning by doing' approach to planning and management is needed. This guidance is helpful in supporting a selective, continuous monitoring approach, linked to an adaptive management style. In this context it would seem that careful definition of economic, social and environmental objectives for marine areas will provide a key mechanism for narrowing the field of data collection to more manageable proportions. Exchange of experience between different marine areas would be particularly beneficial here.

Connecting to wider agendas

Connecting marine and terrestrial planning

An underlying theme in the above discussion is the close relationship between the land and the sea. This reflects the complex interconnectedness of environmental, economic and social systems that operate across terrestrial and marine boundaries.

Both directly and indirectly human development on the land is increasingly having profound impacts on the sea. At the same time, a new appreciation of the dependence of landward communities on the sea for a vast array of ecosystem services is coming to the fore. These extend beyond age old, but still vital, activities such as fishing, to encompass, for example, recognition of the critical part that the sea plays in carbon capture and global climate amelioration. Integration of marine and terrestrial planning and management regimes (including those related to town and regional planning, economic development, energy, water management, agriculture and transport) would therefore appear to be a vital element in the delivery of EA, and, as Chapter 2 discusses, it is evident that a two-way relationship would be beneficial. A particular challenge here will be to encourage greater prominence to be given to environmental matters in many of these land-based plans. For while marine plans are presently being developed largely from an environmental management starting point and often require closer connection to economic and social concerns, the opposite is true for most terrestrial plans. Here, the integration of environmental perspectives in decision-making is in its infancy, despite the widespread adoption in many countries of strategic environmental appraisal and environmental impact assessment. Established political, institutional, professional, and disciplinary norms and practices, let alone legal requirements, can impede effective integration of environmental considerations. There is also a matter of potential 'sea blindness', with many land-based planning institutions not geared up to thinking beyond the coast, and, if they do, to focus upon the sea's impact on the land rather than the impact of land-based activities on the sea.

Challenging the ecological modernization paradigm

In this context, the emergence of EA is partly rooted in an awareness of the failure of established political and administrative systems to respond in an effective way to the environmental challenges we now face. In many countries, a legacy of sectoral environmental management (and other forms of sectoral planning) has encouraged an oversimplification of environmental problems and constrained the development of wider appreciation of the complex and wide-ranging impact of human activity upon the health of ecosystems (Margerum and Born, 1995; Irwin, 2001; Dickens, 2004). The solution, arguably lead by the CBD and EA, is to achieve a transition towards new integrated planning and management structures that use ecological analysis and understanding as a primary basis of decision-making (Harris et al, 1987). Implicit in the CBD interpretation of EA is that humans are an integral part of the natural world and not separate from it and as operational guidance point 2 highlights, application of EA should aim to 'develop a deeper awareness of human/environment connections' through, for example, the concept of ecosystem services. However, one of the criticisms that may be levelled at the CBD approach to EA is that it does not readily offer a sense of clear strategic purpose. While it is firmly rooted in the principles of sustainable development, this term is widely contested and potentially offers a number of very different planning and management approaches. Alternative conceptualizations range from the distinction between 'weak' and 'strong' sustainability as described by Pearce et al (1989) to the 'growth' or 'no-growth' scenarios presented in the Bruntland Report (Bruntland, 1987). The potentially strong divergence in world views that are brought together in planning and managing marine areas are the focus of Chapter 4, which explores current developments in Europe.

Davoudi (2001) argues that despite the variation in definitions available, sustainability discourses can perhaps be best conceptualized in terms of the tension that exists between 'ecological modernization' and 'risk society'. Ecological modernization, she argues, asserts that the environment and the economy are not in conflict and that economic prosperity is necessary to finance measures that promote environmental integrity. Here, the project of modernity can continue to flourish, safe in the belief that science and technology can maintain continuing economic growth and limit environmental damage. 'Risk society' ideas developed by Beck are, however, much more cautious in approach (Beck, 1992). They suggest that it is the by-products of modernity that are responsible for environmental degradation. As a result the 'risk society' view calls not for improved forms of technological development to support continuing economic growth, but for a radical redefinition of the basic norms of society related to business, politics, science and family based around a priority concern for ecosystem (including human) health and well-being. Technological development and continuing economic activity will still feature in this scenario but with a radically different primary goal in mind including the concept of 'fair growth' and even 'no growth' scenarios. Currently, global practice is firmly lodged within the 'ecological modernization' camp. Whether this approach can achieve the ambitions set out in the CBD and EA will remain to be seen. Interestingly though, the new environmentally led marine plans that are emerging, if they can successfully connect with the political, economic and social mainstream, may play a not insignificant role in encouraging debate and reflection upon the current trajectory of human/environment relations.

Conclusions

This chapter began by observing that we are in the midst of a paradigm shift in our approaches to environmental planning and management that reasserts the very ancient idea that human well-being and ecosystem health are closely intertwined. Promoted by the 1992 CBD, the conceptual framework of EA (that reflects this paradigm) has now been embodied in international law and increasingly features in international and national policy documents guiding the new era of planning and management of the sea. However, as we have shown, understanding and application of EA in both marine and terrestrial contexts are still at a relatively early stage, and there have been calls to demystify the concept and develop a clearer understanding of its implications. This book addresses this agenda and this chapter has set the scene for more in-depth examination of related themes in the following chapters by: explaining the origins, definitions and principles of EA and associated UN operational guidance; reviewing some of the lessons that can be drawn from existing experience of applying EA in non-marine and marine areas; and providing a transdisciplinary discussion of key issues that require further attention in developing EA in planning and management of the sea. These include developing understanding of the human dimension of EA (the focus of Chapters 2 and 3), addressing key information challenges presented by the approach (the focus of Chapters 4 and 5), and exploring how EA in marine planning and management might connect more closely with wider public policy and research agendas (the focus of Chapter 6).

References

Allen, D. E. (1976) *The Naturalist in Britain: A Social History*, Pelican Books, London

Backer, H. and Leppanen, J. M. (2008) 'The HELCOM system of a vision, strategic goals and ecological objectives: Implementing an ecosystem approach to the management of human activities in the Baltic Sea', *Aquatic Conservation-Marine and Freshwater Ecosystems*, vol 18, pp321–334

Backer, H., Leppänen, J.-M., Brusendorff, A.C., Forsius, K., Stankiewicz, M., Mehtonen, J., Pyhälä, M., Laamanen, M., Paulomäki, H., Vlasov, N. and Haaranen, T. (2009) 'HELCOM Baltic Sea Action Plan – A regional programme of measures for the marine environment based on the Ecosystem Approach', *Marine Pollution Bulletin*, vol 60, no 5, pp642–649

Beck, U. (1992) *Risk Society: Towards a New Modernity*, Newbury Park, London

Bengston, D., Xu, G. and Fan, D. (2001) 'Attitudes toward ecosystem management in the United States, 1992–1998', *Society & Natural Resources*, vol 14, no 6, pp471–487

Bruntland, G. (ed) (1987) *Our Common Future: The World Commission on Environment and Development*, Oxford University Press, Oxford

CBD COP (Convention on Biological Diversity, Conference of the Parties) (2000) *Fifth Meeting, Decision V/6 The Ecosystem Approach*, Secretariat of the Convention on Biological Diversity, Montreal

CEC (Commission of the European Communities) (1996) *Demonstration Programme on Integrated Management of Coastal Zones, European Commission Services, Information Document XI/79/96*, Office for the Official Publications of the European Communities, Luxembourg

CEC (2007) *An Integrated Maritime Policy for the European Union*, Office for the Official Publications of the European Communities, Luxembourg

CONSSO (2002) *Bergen Declaration*, Ministerial declaration of the fifth international conference on the protection for the North Sea, 20–21 March 2002, CONSSO, Bergen

Constable, A. J., de la Mare, W. K., Agnew, D. J., Everson, I. and Miller, D. (2000) 'Managing fisheries to conserve the Antarctic marine ecosystem: Practical implementation of the Convention on the Conservation of Antarctic Marine Living Resources (CCAMLR)', *Ices Journal of Marine Science*, 57 pp78–79

Danter, K. J., Griest, D. L., Mullins, G. W. and Norland, E. (2000) 'Organizational change as a component of ecosystem management', *Society & Natural Resources*, vol 13, no 6, pp537–548

Davoudi, S. (2001) 'Planning and the twin discourses of sustainability', in A. Layard, *Planning for a Sustainable Future*, Spon, London

Day, J. (2008) 'The need and practice of monitoring, evaluating and adapting marine planning and management: Lessons from the Great Barrier Reef', *Marine Policy*, vol 32, pp823–831

Defra (2009) *Our Seas: A Shared Resource – High Level Marine Objectives*, Defra, London

Dickens, P. (2004) *Society and Nature*, Polity, Cambridge

Endter-Wada, J., Blahna, D., Krannich, R. and Brunson, M. (1998) 'A framework for understanding social science contributions to ecosystem management', *Ecological Applications*, vol 8, no 3, pp891–904

Fee, E., Gerber, E., Rust, J., Haggenmueller, K., Korn, H. and Ibiscf, P. (2009) 'Stuck in the clouds: Bringing the CBD's ecosystem approach for conservation management down to Earth in Canada and Germany', *Journal for Nature Conservation*, vol 17, no 4, pp212–227

Frid, C., Paramor, O. and Scott, C. (2006) 'Ecosystem-based management of fisheries: Is science limiting?', *ICES Journal of Marine Science*, vol 63, no 9, pp1567–1572

Glowka, L., Burhenne-Guilmin, F. and Synge, H. (1994) *A Guide to the Convention on Biological Diversity*, IUCN, Cambridge

Harris, H. J., Sager, P. E., Yarbrough, C. J. and Day, H. J. (1987) 'Evolution of water resource management: A Laurentian Great Lakes case study', *International Journal of Environmental Studies*, vol 29, no 1, pp53–70

Hartig, J. H., Zarull A. Z. and Law, N. L. (1998) 'An ecosystem approach to Great Lakes management: Practical steps', *Journal of Great Lakes Research*, vol, 24, no 3, pp739–750

Hewitt, R. P. and Low, E. H. (2000) 'The fishery on Antarctic krill: Defining an ecosystem approach to management', *Reviews in Fisheries Science*, vol 8, no 3, pp235–298

Hillman, M., Aplin, G. and Brierley, G. (2003) 'The importance of process in ecosystem management: Lessons from the Lachlan catchment, New South Wales, Australia', *Journal of Environmental Planning and Management*, vol 46, no 2, pp219–237

Holling, C. S. (1978) *Adaptive Environmental Assessment and Management*, Wiley, Chichester

Holling, C. S. (1993) 'Resilience and stability of ecological systems', *Annual review of Ecological Systems*, vol 4, pp1–23

Holling, C. S. (2001) 'Understanding the complexity of economic, ecological, and social systems', *Ecosystems*, vol 4, pp390–405

Hyun, K. (2005) 'Transboundary solutions to environmental problems in the Gulf of California Large Marine Ecosystem', *Coastal Management*, vol 33, no 4, pp435–445

Interagency Ocean Policy Task Force (2009) *Interim Framework for Effective Coastal and Marine Spatial Planning*, White House Council on Environmental Quality, Washington, DC

Intergovernmental Oceanographic Commission (2007) National Ocean Policy, The Basic Texts from: Australia, Brazil, Canada, China, Colombia, Japan, Norway, Portugal, Russian Federation, United States of America. UNESCO, Paris

International Institute for Sustainable Development (2006) *Earth Negotiations Bulletin, vol 25, no 31, Summary of the Seventh Meeting of the Open-ended Informal Consultative Process on Oceans and the Law of the Sea*, IISD, Winnipeg

Irwin, A. (2001) *Sociology and the Environment*, Polity, Cambridge

Juda, L. (1999) 'Considerations in developing a functional approach to the governance of large marine ecosystems', *Ocean Development & International Law*, vol 30, no 2, pp89–125

Kidd, S. (2007) 'Landscape planning at the regional scale', in M. Roe (ed) *Landscape and Sustainability*, 2nd edition, Spon, Abingdon, pp118–137

Klug, H. (2002) 'Straining the law: Conflicting legal premises and the governance of aquatic resources', *Society & Natural Resources*, vol 15, no 8, pp693–707

Laffoley, D. d'A., Maltby, E., Vincent, M. A., Mee, L., Dunn, E., Gilliland, P., Hamer, J. P., Mortimer, D. and Pound, D. (2004) *The Ecosystem Approach: Coherent Actions for Marine and Coastal Environments*, English Nature, Peterborough

Lamont. A. (2006) 'Policy characterization of ecosystem management', *Environmental Monitoring and Assessment*, vol 113, pp5–18

Lindblom, C. (1959) 'The science of muddling through', *Public Administration Review*, (reprinted in A. Faludi (1973) *A Reader in Planning Theory*, Pergamon, Oxford)

Maes, F. (2008) 'The international legal framework for marine spatial planning', *Marine Policy*, vol 32, pp797–810

Maltby, E. (2006) 'Wetland conservation and management: Questions for science and society in applying the ecosystem approach', *Ecological Studies*, vol 191, pp93–115

Maltby, E., Holdgate, M., Acreman, M. and Weir, A (1999) *Ecosystem Management: Questions for Science and Society*, IUCN CEM and RHIER, London

Margerum, R. and Born, S. (1995) 'Integrated environmental management: Moving from theory to practice', *Journal of Environmental Planning and Management*, vol 38, pp371–391

Marten, G. (2001) *Human Ecology*, Earthscan, London

McLoughlin, B. (1969) *Urban and Regional Planning: A Systems Approach*, Faber & Faber, London

More, T. (1996) 'Forestry's fuzzy concepts: An examination of ecosystem management', *Journal of Forestry*, vol 94, no 19, pp23–24

MSPP Consortium (2006) *Marine Spatial Planning Pilot*, MSPP, London

Pearce D. W., Markandya, A. and Barbier, E. B. (1989) *Blueprint for a Green Economy*, Earthscan, London

SBSTTA (Subsidiary Body on Scientific, Technical and Technological Advice) (2007) *In-depth Review of the Application of the Ecosystem Approach*, UNEP/CBD/SBSTTA/12/2, Secretariat of the CBD, Montreal

Scoones, I. (1999) 'New ecology and the social sciences: What prospects for fruitful engagement?', *Annual Review of Anthropology*, vol 28, pp479–507

Secretariat of the CBD (Convention on Biological Diversity) (2010a) 'Ecosystem Approach', Secretariat of the CBD, Montreal www.cbd.int/ecosystem, accessed 20 June 2010

Secretariat of the CBD (2010b) 'Ecosystem Approach Source Book', Secretariat of the CBD, Montreal, www.cbd.int/ecosystem/sourcebook, accessed 20 June 2010

Shepherd, J. (2008) *The Ecosystem Approach: Learning from Experience*, IUCN, Gland, Switzerland and Cambridge

Smith, R. D. and Maltby, E. (2003) *Using the Ecosystem Approach to Implement the Convention on Biological Diversity: Key Issues and Case Studies*, IUCN, Gland, Switzerland and Cambridge

The Swiss Agency for the Environment, Forests and Landscape, the Bureau of the Convention on Wetlands and the World Wide Fund for Nature (2002) *Sustainable Management of Water Resources: The Need for a Holistic Ecosystem Approach*, Ramsar COP8 DOC. 32, Secretariat to the Ramsar Convention, Gland, Switzerland and Cambridge

UNEP (United Nations Environment Programme) (2002) *Global Environmental Outlook 3*, Earthscan, London

UNEP (2010) *Regional Seas Programme*, www.unep.org/regionalseas/about/default.asp, accessed 20 April 2010

Wang H. (2004) 'An evaluation of the modular approach to the assessment and management of large marine ecosystems', Ocean Development and International Law, vol 35, pp267–286

Chapter 2

Developing the Human Dimension of the Ecosystem Approach: Connecting to Spatial Planning for the Land

Sue Kidd, Jim Claydon, Nigel Watson and Stuart Rogers

This chapter aims to illustrate:

- Why close connection to spatial planning for the land may be of value in developing the human dimension of marine planning and management;
- How key theoretical debates regarding the purpose and process of terrestrial planning might serve to stimulate critical reflection upon the purpose and process of planning for the sea;
- How new marine planning and management regimes can draw upon terrestrial planning experience and be tailored to particular cultural, legal and administrative contexts; and
- How terrestrial planning can assist in understanding and delivering key aspects of EA.

Introduction

Chapter 1 introduced the ecosystem approach (EA) as the prevailing framework guiding the development of new approaches to marine planning and management at the present time. However, it highlighted that the application of EA within the marine context poses significant challenges and that a key area in need of further development is the human dimension that is so central to the approach. For example, in line with the holistic perspective of EA, more prominent recognition of the close interrelationship between humans and the environment and between terrestrial and marine areas is required than is so often the case at the present time. Equally, EA emphasizes that managing human activity, rather than the environment itself, must inevitably be the main focus of attention. This is particularly so in the marine context where we cannot even attempt to manage the complexities of the wider ecosystem. Following this logic it is therefore not surprising that Principle 1, put forward by the Conference of the Parties (COP) to the Convention on Biological Diversity (CBD) in May 2000, states that the objectives of management of land, water and living resources are a matter of societal choice, as they inevitably will entail directing and potentially limiting human activity in some way. Such an assessment suggests that it is important that robust mechanisms are put in place to clearly determine what social, economic and environmental ambitions for the sea actually are, and to tease out the key characteristics of the future marine environment that should be strived for. If this is accepted, then the process of marine planning and management needs to be designed in a way that not only incorporates good science but also, as Principle 12 indicates, effective engagement with all relevant sectors of society. This is an ambitious agenda for those building new marine planning and management regimes at the present time. However, this chapter argues that they should take heart in the fact that this situation is not entirely new and that much can be gained by reflecting on the experience of terrestrial planning which has been grappling with similar challenges for more than 100 years.

In her paper in the special issue of the journal *Marine Policy* on marine spatial planning (MSP) Fanny Douvere (2008) also highlighted the value of reflecting upon terrestrial planning experience and noted that the potential benefits of this connection are now widely recognized. For example, this is evident in an initiative by the Intergovernmental Oceanographic Commission (IOC) and the Man and the Biosphere Programme of the United Nations Educational, Scientific and Cultural Organization (UNESCO) which drew together existing experience of managing the multiple use of marine space and spatial planning experience from the land (Ehler and Douvere, 2007). Quite rightly, much of the work to date has focused upon *current* land-based spatial planning practice and how this can inform the development of marine plans. Some of the key outputs from this work are discussed in Chapter 5. However, less attention has so far been given to historical reflection and discussion of underlying theoretical debates that have informed changing views as to the purpose and process of land-based spatial planning over the years and how this might equally inform the emerging field of marine planning. This chapter aims to illustrate the richness and value of potential linkages here and encourage ongoing theoretical and practice debate and exchange across the land/sea divide.

The chapter starts by outlining the origins of planning for the land and the striking similarities and also differences to the current context in which more formalized planning and management for marine areas is emerging. It then provides a selected potted history of key theoretical debates regarding the purpose and process of terrestrial planning and shows, through the use of examples, how this might serve to stimulate critical reflection upon the purpose and process of planning for the sea. Much of the discussion relates to the content of plans and the process of plan-making, but attention is also drawn to the critical issue of implementation which, it is argued, merits greater prominence in marine planning and management discussions at the present time. This section concludes with consideration of the current 'spatial planning' paradigm that is increasingly guiding approaches to land-based planning around the world, but also with a reminder that planning arrangements need to be carefully tailored to particular cultural, legal and administrative contexts. The practical application of this type of understanding is then illustrated by reference to the new marine spatial planning arrangements that have been established in the UK under the 2009 Marine and Coastal Access Act. These have been informed in a very direct way by the UK's terrestrial planning arrangements. Finally the chapter ends with some conclusions about how terrestrial planning experience can assist more generally in understanding and delivering key aspects of EA.

Common foundations

The history of planning on the land goes back many millennia. Some of the earliest examples can be found in the ancient cities of Mesopotamia, 4500 years ago, and the planned settlements of the ancient Greeks and Romans (Mumford, 1961). The development of modern planning systems for the land is, however, of more recent origin. It surfaced as a response to the linked issues of industrialization, population growth and urban expansion that were a feature of Western Europe and North America in the latter part of the 19th century.

'The City of the Dreadful Night' (Hall, 2002, p13), the title of a poem by Victorian poet James Thompson, evocatively conjures up some of the concerns of the day. These included increasing awareness of the terrible living conditions that millions of urban dwellers were experiencing in the rapidly built, unplanned and unregulated 19th century cities. Problems of overcrowding, lack of light and sanitation and associated impacts in the form of poor health and low life expectancy were highlighted in a series of urban surveys of the period which attracted widespread popular attention (Cullingworth and Nadin, 2006). Such altruistic social concerns were also paralleled by issues of national security, with fears of violence and insurrection fuelled by the experience of popular uprisings that had been a recurring feature in Europe throughout the 18th and 19th centuries. These conditions had resulted in a flight from the cities of the better-off middle classes, and the start of the process of mass suburbanization, a particular feature of the UK and the US. George Cruikshank's cartoon published in 1829 (see Figure 2.1), depicting the outward spread of London, illustrates the

Figure 2.1 'London Going Out of Town' or 'The March of Bricks and Mortar'

Source: An etching by George Cruikshank from 1829. By permission of the Museum of London

intensity of sentiments aroused by increasing urban encroachment on the countryside and associated fears about the adverse impacts upon wildlife, natural beauty and agricultural production (Gallent et al, 2008). It is within this context that public support grew for the introduction of measures to regulate development in order to prevent some of the adverse social and environmental impacts of urban growth. These culminated in the UK, for example, in the passing in 1909 of the first formal town planning legislation – the Housing, Town Planning etc. Act.

One hundred years later, it is interesting to compare this experience with the concerns that are now being voiced in support of better controls over human use of the sea. A key document here is the Millennium Ecosystem Assessment for marine and coastal ecosystems (UNEP, 2006) produced by the United Nations Environment Programme (UNEP). The report highlights the importance of ocean and coastal ecosystems to human survival and well-being. However, it concludes that these ecosystems are being degraded and used unsustainably and are deteriorating faster than other ecosystems. Population growth, technological change and shifting consumer demands are all placing increasing demands and threats on the marine environment. These include threatened food security, loss of habitat, adverse health impacts and increased vulnerability of coastal communities to natural and human-induced disasters. Such concerns, related particularly to the adverse social and environmental impacts of unregulated human use of the sea, closely mirror those that led to the establishment of land-based planning systems in many parts of the world a century ago. However, today, economic justifications for planning intervention are also to the fore, although often couched in the language of sustainable development. These concerns were not so prominent many years ago but they are, for example, a feature in current European debates about *Maritime* Spatial Planning, revealing a subtle but potentially significant distinction from the arguably more environmentally orientated *Marine* Spatial Planning. A quote from the Commission of the European Communities (CEC) illustrates the difference in emphasis.

> *Between 3 and 5% of Europe's Gross Domestic Product (GDP) is estimated to be generated by maritime industries and services, some with high growth potential. A stable planning framework providing legal certainty and predictability will promote investment in such sectors, which include offshore energy development, shipping and maritime transport, ports development, oil and gas exploitation and aquaculture, boosting Europe's capacity to attract foreign investment. (CEC, 2008a, p3)*

It is therefore clear that a range of powerful interests are coalescing, perhaps a little uneasily, to press the case for better guidance and regulation of human development of the sea. As a consequence it can be anticipated that, during the first quarter of the 21st century, governments across the world will be considering what form such an approach might take. This chapter argues that, in undertaking these activities, they will find it helpful to look at the planning regimes that they have developed for the land as they share a common foundation and concern. As will be illustrated below, an appreciation of how they have changed over time in response to shifting perceptions as to the purpose and process of planning opens up many useful lines of enquiry.

Considering the purpose of marine planning and management

The concerns leading to the establishment of planning regimes for the land and the sea give some clues as to varying perspectives on the intended purpose of planning. This has been a significant area of debate in terrestrial planning during the past century. The importance of this type of reflection is highlighted by the fact that distinct differences in emphasis are evident among groupings and over time, indicating that the purpose of planning is far from straightforward or value-free. Varying viewpoints reflect different cultural and political perspectives which are in themselves in constant flux. Ultimately all planning regimes involve implementing a particular normative position – a set of values that are dominant at a particular point in time which inform the setting of objectives and also shape the processes and techniques used to pursue those objectives. Appreciation of this reveals the importance of ongoing discussion about the purpose of any planning activity.

Within the terrestrial context, much of the debate has centred upon the kinds of environment that planning activity should seek to create and protect. There has been a particular emphasis upon the urban environment here, in part reflecting the fact that this is where most people live, most development takes place, where most competition and potential conflicts arise, and therefore where planning is most needed. Starkly different views of the 'ideal' urban environment have held sway at different points in time. It is impossible here to do justice to the depth of discussion that has taken place, but Table 2.1 illustrates some of the key contrasts in perspective that have been evident and how mainstream thinking has shifted over the years.

In the early part of the 20th century, attention was focused mainly upon the physical aspects of towns and cities and the development of universally applicable 'blueprint' visions of orderly, often low density, comprehensively planned settlements where land uses were carefully zoned and segregated in different ways. This approach is reflected for example in Ebenezer Howard's 'garden cities', Le Corbusier's 'cities in the sky' and Frank Lloyd Wright and Lewis Mumford's 'decentralized cities' (see Hall, 2002, for an excellent overview). However, by the second half of the 20th century, these perspectives were facing increasing criticism. Lead by authors such as Jane Jacobs (1961) Christopher Alexander (1965) and Brian McLoughlin (1969) they were accused as being overly simplistic and sterile, and lacking in appreciation of the intricate and

Table 2.1 *Dominant planning views on the 'ideal' urban environment*

Early 20th century	Late 20th century	Early 21st century
Focus upon physical dimensions	Focus on social/economic dimensions	Integration of ecological/ social/economic dimensions
Order and simplicity	Complexity and richness	Respecting natural capacities
Comprehensively planned	Incrementally changing	Adaptable
Ideal end state (blueprint)	Dynamic	Resilient
Dispersed development	Compact development	Compact development
Low density development	High density development	High density development
Segregation of uses	Mixed use	Multifunctional
Self-contained communities	Interconnected communities	Resource efficient
Universal values	Diverse values	Shared responsibilities

Changing planning paradigms

$\longrightarrow$

multifaceted social and economic life of settlements which many regarded as their key attraction and value. Instead, the complexity and dynamic nature of high density cities were seen as attributes to be understood, valued and nurtured by planning activity, not only because of their aesthetic and intellectual interest, but also because these features were perceived as key underpinnings of social well-being and economic success (see Taylor, 1998, for a useful summary of these changing perspectives). More recently, traditional support for carefully zoned, low density development has also been criticized on sustainability grounds, not least because of resource concerns related to minimizing the need for greenfield development, but also because research has indicated that such settlement patterns tend to increase the need to travel and are therefore more energy intensive (Newman and Kenworthy, 1999).

Since the Rio Earth Summit of 1992, this concern for sustainability has increasingly become the central focus of planning debate. It is reflected, for example, in the UK, where the Planning and Compulsory Purchase Act of 2004 defined for the first time a statutory purpose for terrestrial planning which is to promote sustainable development. Drawing inspiration from the Bruntland report (World Commission on Environment and Development, 1987), considerable attention has been directed towards distinguishing those environmental qualities that are consistent with ecological integrity, social equity, health and well-being and economic prosperity and fostering an appropriate balance between them. This shift in thinking has altered professional and public expectations of planning itself and has prompted a revival of interest in urban design and physical planning reflected in the new urbanism (Katz, 1994), smart growth (Ingram et al, 2009) and urban village (Neal, 2003) movements as well as further development of complex city ideas (see, for example, Frey and Yaneske, 2007). More recently, mirroring the intensity of international interest surrounding the December 2009 United Nations (UN) Climate Change Conference in Copenhagen, environmental dimensions and shared global responsibilities have come to the fore. There is in particular a new sense of urgency related to the promotion of carbon conscious forms of development (ISOCARP, 2009) that mitigate against and adapt to climate change. Current thinking proposes that we should focus on creating high density, compact, adaptable and resource efficient settlements with an emphasis on multifunctionality and sensitivity to ecological processes and natural carrying capacity. It is salutary to note that in many respects these ideas are not entirely new and resonate for example with the bioregional planning concepts put forward by Patrick Geddes, Lewis Mumford and others in the early years of the 20th century (Luccarelli, 1995). What we now see therefore is an interesting coming together of longstanding elements of planning thought, about the ideal urban environment, recast and developed to address current understanding of the key challenges of the day. Such paradigm shifts regarding planning outcomes and aspirations clearly have significant implications for the types of planning processes, techniques and skills deemed to be most effective and desirable.

So what can marine planning and management learn from this experience? There is much to reflect on here and three examples serve to illustrate the potential value in exploring the linkages further. Firstly, it is interesting to observe that much terrestrial planning attention has been concerned with defining the 'ideal' attributes of those areas of most intense development – our towns and cities. A similar approach would seem to be relevant for the sea and suggests a case for particular discussion around how to plan for the continental shelves and, more specifically, inshore 'urban sea' areas where human pressures are at their greatest and interactions with land-based development are most pronounced. These areas are also critical in terms of marine ecosystem functioning given that most life in the sea depends upon them, either directly or indirectly. They therefore potentially require a different and more active type of planning response to more rural and wilderness seas and deep ocean areas and teasing out these distinctions, subtleties and also interrelationships could be helpful.

Secondly, there is a question about how best to guide human activity, or, in other words, are there 'ideal' planning strategies for the development of the sea? Experience from the land suggests the need to be wary of simplistic or 'one size fits all' solutions and to carefully tailor responses to particular localities, respecting and working with system complexity in a sensitive manner. In this context, the emphasis that is currently being placed on ocean zoning in some quarters as a core marine spatial planning mechanism, where particular sectors with defined needs are allocated sea-space for their exclusive or near-exclusive use, needs critical examination. While there is no doubt it has a place, experience suggests that it is possible for many sea uses to live happily alongside each other if appropriate care is taken, and that simplistic approaches to zoning may provoke unnecessary competition and conflict between users. These could, for example, make the enforceability of zoning controls problematic, or have negative and unforeseen consequences such has diverting human pressure to other areas where controls are not in force. This phenomenon, sometimes called leapfrogging, has been well charted in relation to green belt and other planning controls in many parts of the world. Indeed the scope for multifunctionality within a single region of the sea may be greater than that of the land as it experiences more evident temporal variations (daily, monthly, seasonal, and so on), activities may take place in different parts of the water column and may be much less fixed to a specific location than is typically the case with landward development, so the scope for accommodation may be larger. The relative stability of the marine environment in comparison to the terrestrial environment also supports this view. Nurturing multifunctionality through the development of criteria based planning policies and case-by-case decision-making and tailored controls could be a key feature of planning for sea, with use segregation through zoning perhaps playing a lesser role than is sometimes the case on the land. Healthy debate around these issues certainly seems to be desirable. That does not mean that there is no value in also thinking about the merits of low density versus high density and dispersed versus compact forms of development of the sea and other potential spatial variants. There seems to be a critical need to move beyond simple statements of high level objectives (such as those illustrated in Box 2.1) to develop a more precise understanding of what sustainable patterns of sea use might look like. Intuitively, the current thinking favouring high density compact forms of development on the land for social, economic and environmental reasons seems to be equally applicable to the sea, but this also needs critical examination.

Thirdly, and following on from this, arguably one of the most important considerations affecting views as to the 'ideal' marine environment of the future will be climate change, just as it is becoming crucial to determining the future patterns of development on the land. However, are we clear what carbon conscious planning in a world facing rapid population growth might mean for the patterns of human development in the sea? It could well suggest quite different objectives for and approaches to marine planning and management than has been the case to date. It certainly seems to require a forward looking, plan-led approach in which the inevitability of biophysical and human change

---

Box 2.1 *Examples of high level objectives related to the purpose of marine planning and management*

'Promote the peaceful use of the seas and oceans, the equitable and efficient utilization of their resources, the conservation of their living resources, and the study, protection and preservation of the marine environment.' (UN, 1982, UNCLOS preamble)

'Support sustainable, safe secure, efficient, and productive uses of the ocean, our coasts and the Great Lakes, including those that contribute to the economy, commerce, recreation, conservation, homeland and national security, human health, safety and welfare.' (White House Council on Environmental Quality, 2009, p7)

'The marine environment is a precious heritage that must be protected, preserved and, where practicable, restored with the ultimate aim of maintaining biodiversity and providing diverse and dynamic oceans and seas which are clean, healthy and productive.' (CEC, 2008b, para 3)

'Clean, healthy, safe, productive and biologically diverse oceans and seas.' (Defra, 2009, p3)

needs to be actively recognized and embraced. While established concerns to preserve and even restore the ecological integrity of marine areas may intensify under such a scenario given the significant carbon capture role of marine ecosystems, it is also likely that many will see much more active development of marine areas as critical to a low carbon future. For example, calls for expansion of marine-related renewable energy generation, marine aquaculture and carbon storage below the seabed are mounting, as is evident in the quotation from the EU's Maritime Policy referred to above. The difficulties of achieving the change needed to mitigate and adapt to climate change, suggest that planning for the land and the sea will inevitably be closely interwoven over the coming years. In this respect it can be envisaged that views related to the 'ideal' qualities of the landward environment (for example, public attitudes to the location of wind farms) may well significantly influence opinions as to the desirable uses and qualities of the marine environment. The development of our understanding of what might be meant by low carbon, zero carbon or even carbon positive development is therefore likely to be important for the sea, just as it is in relation to landward development.

Developing the process of marine planning and management

Mirroring theoretical debates regarding the purpose of terrestrial planning has been an even larger body of work related to the planning process. Again, distinct differences in approach are evident over time and it is impossible to do justice to the richness of the debate in the chapter. However, Table 2.2 provides an overview of key shifts in thinking that are of particular interest to marine planning and management.

As we have seen, during the early years of the modern planning movement, and reflecting historical traditions dating back to the Romans and Ancient Greeks and before, terrestrial planning activity was regarded primarily as a process of physical design. Architecture and town planning were closely linked, with the developing 'art' of town planning extending aesthetic appreciation to

Table 2.2 *Dominant views on the 'ideal' planning process*

Early 20th century	Mid-20th century	Late 20th century	Early 21st century
Planning as a design process	Planning as a scientific process	Planning as a communicative process	Spatial planning
	Changing planning paradigms		

the design and remodelling of whole towns. However, by the 1960s, this view was subject to growing criticism not only related to the limitations of the environments that were being created, but also to the processes through which plans were devised. Brian McLoughlin was a leader in the movement to a more rational approach to planning that drew upon scientific traditions and reflected developments in cybernetics and ecology of the period. In his famous book *Urban and Regional Planning: A Systems Approach* (McLoughlin, 1969) he proposed that urban areas and regions should be considered as complex systems of activities and places that were interrelated and dynamic, with change in one area having impacts elsewhere. As a consequence, McLoughlin believed that planning should entail a rational and ongoing process of systems analysis and control, with the focus upon understanding the dynamics of urban areas through extensive information gathering and modelling, and the development of adaptable and flexible plans that steered development in a desired direction. This was a far cry from the blueprint, comprehensive redevelopment and design-based thinking of the early years of planning. This notion of the planning process as a science rather than an art took hold and, although it did not completely replace artistic traditions, it was widely taken on board by terrestrial planners, especially those working at more strategic scales, and it remains influential today. So developed a model of the planning process as a logical progression through; survey, analysis, consideration of alternatives, plan finalization, implementation, monitoring and review. Such thinking has, however, in turn been subject to rigorous and wide-ranging critique. Most fundamentally, this reflects the underlying shift in philosophical thought from modernism to postmodernism which has questioned the very notion of impartial rationality. However, before we move on to discuss this it is worth drawing attention to one particular line of debate within the rational planning paradigm itself which has resonance with current discussions in marine planning and management.

Consequent upon a rational process view of planning is an ambition for comprehensiveness. In order to make rational decisions, detailed information about the many facets of the urban (or marine) system must be assembled and carefully analysed, and the consequences and merits of possible alternative courses of action should be assessed in a rigorous manner. However, as Charles Lindblom (1959) and many others have argued, the practicalities of most planning situations mean that time and resources are limited and ambitions for rationality and comprehensiveness are, more often than not, replaced by a piecemeal, incremental, opportunistic and pragmatic process that Lindblom called 'disjointed incrementalism'. One response to Lindblom's analysis was the idea of mixed scanning put forward by Amitai Etzioni (1967). This entails, for example, distinguishing strategic decisions from those of a more detailed and operational nature. This raises questions regarding the 'level' or 'levels' at which marine spatial planning should be developed and the kinds of planning processes, data and skills needed to enable its effective functioning at the selected level(s).

However, alternative schools of thought have emerged that question the relevance and appropriateness of both the rational-comprehensive and incremental approaches to planning. In particular, theories of complexity have been used to emphasize the dynamic, open-ended nature of environmental systems and the inherent uncertainty that surrounds the 'wicked' and 'messy' socio-biophysical problems they generate. Such debates draw attention to the limitations of scientific knowledge and the impossibility of accurate predictions of future conditions and responses (Trist, 1980; Dryzek, 1997). As such, the search for rationality and comprehensiveness in many areas of spatial planning (which can be considered to include both terrestrial and marine planning) may be unrealistic or at the very least bounded, implying that planning styles geared more towards trial-and-error experimentation, controlled risk-taking and long-term adaptation may be more appropriate in an 'age of uncertainty' (Christensen, 1985; Briassoulis, 1989). Interestingly, these conclusions are reflected in EA with Principle 9 emphasizing the importance of recognizing that change is inevitable and operational guidance point 3 promoting adaptive management approaches.

Although apparently presenting distinctly opposing ideas about the nature of the planning process, the art and science perspectives discussed above do, however, have something in common. They share a view of the planner as an expert who, through the application of specialist knowledge and skill, is able to make judgements on behalf of the general public about what kinds of environment it is desirable to create. In line with the wider attack on modernism and the development of postmodern lines of thought, this notion of an impartial or value-free planning process has increasingly been challenged. For example, it is argued that, unlike traditional concepts of science, planning is not simply a descriptive process concerned with explaining the world. It is also a prescriptive activity focused upon intervening in some way in the normal course of things to protect or improve the environment with some purpose in mind. Taylor summarizes the critique very well.

> *The prime questions facing any planning exercise are questions about what it is best to do, and these are questions of value. Town planning is not a science in the usual sense of that word. It is therefore more accurate to define planning as an evaluative or normative activity. Moreover, since town planning can significantly affect the lives of large numbers of people, and since different individuals and groups may hold different views about how the environment should be planned, based on different values and interests, it is therefore also a political activity. (Taylor, 1998, p83)*

Consequent upon this analysis, during the late 20th century a new view of the planner and the planning process emerged. This has not replaced but rather runs alongside, and arguably moderates, the art and science perspectives described above. This sees the planner as a communicator and mediator between different interests, and planning as a transactive process aimed at fostering communicative or collaborative action. Drawing upon Habermas' theory of communicative action (1984), and developed within a planning context by authors such as Healey (1997) and Innes and Booher (1999), this approach acknowledges the existence of multiple viewpoints and stresses the importance of consensus building in developing equitable, robust and implementable planning strategies. Arguments in favour of this approach have not just been confined to the growing acknowledgement of the existence of very different perspectives regarding planning interventions. They have also arisen from increasing appreciation of the complexity of current planning issues, such as urban regeneration, climate change, river basin management and sustainable development, which have highlighted the need for joined-up or integrated action across traditional sectoral and territorial divides (Cowell and Martin, 2003; Watson, 2004). The process sometimes referred to as the 'hollowing out' of the state has been an additional factor promoting this view in some countries. This has seen a shift in the role of the state from being a direct provider of services to that of an enabler which increasingly steers, commissions or encourages others to work on its behalf in a more complex arrangement of 'governance'. Close engagement of these parties in decision-making is therefore critical to the delivery of public policy objectives. In this context, facilitation, communication, multi-party collaboration and listening skills are now acknowledged as key attributes for planners. Although well-embedded in many terrestrial planning systems across the world, this communicative view of planning has been criticized as being idealistic, potentially misleading and even naive. The core issues here relate to the exercise of power within the decision-making process and the improbability of achieving true consensus. For example, collaborative approaches do not necessarily empower the weakest voices and may only serve to increase the influence of non-elected powerful interest groups (Watson et al, 2009). Similarly, control of the planning process in most instances firmly remains with the state and the political powers that hold sway. Some argue that this is proper in democratically elected systems, but equally achieving openness and genuine consensus is difficult in top-down and bureaucratic institutional structures. As such, collaborative planning appears to offer certain advantages but in reality may prove to be no better, or even worse, in terms of fostering agreement and action than traditional expert-led exercises.

As terrestrial planning has matured, such debates have continually shaped and reshaped the process of planning and the latest paradigm that holds sway in

Figure 2.2 *Planning process traditions encompassed within spatial planning*

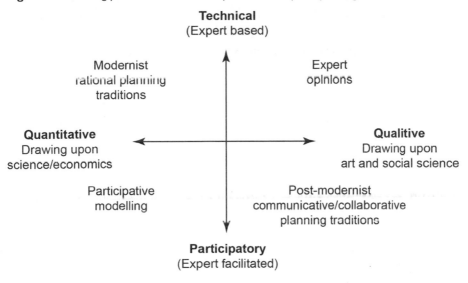

many parts of the world – *spatial planning* – can be seen to represent an amalgam and refinement of much that has gone before (see Figure 2.2).

One of the earliest definitions of spatial planning was put forward in the European Regional/Spatial Planning Charter of 1983.

> *Regional/spatial planning gives geographical expression to the economic, social, cultural and ecological policies of society. It is at the same time a scientific discipline, an administrative technique and a policy developed as an interdisciplinary and comprehensive approach directed towards a balanced regional development and the physical organization of space according to an overall strategy. (Committee of Ministers to Member States on the European Regional/Spatial Planning Charter, 1983, Recommendation 84(2))*

This and other definitions of spatial planning tend to emphasize two central features that weave together the science, design and communicative views of the planning process. (Tewdwr-Jones and Williams, 2001; Healey, 2004; Schön, 2005). The first is a focus upon understanding and guiding the spatial organization of particular places to achieve agreed economic, social, cultural and ecological ambitions. This encompasses a concern, for example, with the distribution, quality and compatibility with current and future needs and aspirations for physically static elements (such as natural resources, housing, employment uses, key infrastructure, community facilities etc.) and also for movement and flows (economic, social, environmental and physical) between different uses and areas. Particularly at more strategic levels, spatial planning therefore carries forward, at least to some degree, a scientific, systems-influenced and evidence-based approach to planning. At a more local level, it also supports a continuing and even reinvigorated interest in urban design and the 'art' of place making. The second feature of spatial planning relates to a concern to achieve policy coherence and support for planning interventions among the variety of authorities and stakeholders that influence spatial development patterns in some way. In this way, spatial planning also takes on board communicative and collaborative planning ideas. The search for integration is common to both these features and, as Table 2. 3 illustrates, it includes a number of dimensions that help to shape all aspects of the planning process, from the earliest stages of plan scoping and formulation through to implementation, monitoring and review.

Figure 2.3 sets out the formal process through which the new breed of local spatial plans are being prepared in England. This clearly shows the emphasis being placed upon evidence gathering and a staged, rational approach to decision-making. At the same time, in line with communicative and collaborative planning perspectives, community involvement is highlighted as an essential feature throughout. So too is the notion of close monitoring of wider contextual and plan-related output indicators (ODPM, 2005). This is not just a feature within the plan development period, but also subsequently in the implementation phase, in order to inform the need for plan adjustment and review.

Table 2.3 *A framework for integration in spatial planning*

Sectoral	Cross-sectoral integration	Integration of different public policy domains
	Inter-agency integration	Integration of public, private and voluntary sector activity
Territorial	Vertical integration	Integration between different scales of planning and management activity
	Horizontal integration	Integration of planning and management activity between adjoining areas or areas with some shared interest
Organizational	Strategic integration	Integration of planning and management strategies, programmes and initiatives
	Operational integration	Integration of delivery mechanisms in all relevant agencies
	Disciplinary/stakeholder integration	Integration of different disciplines and stakeholders

Source: Kidd (2007, p167)

Figure 2.3 *Summary of the process for preparing English development plan documents*

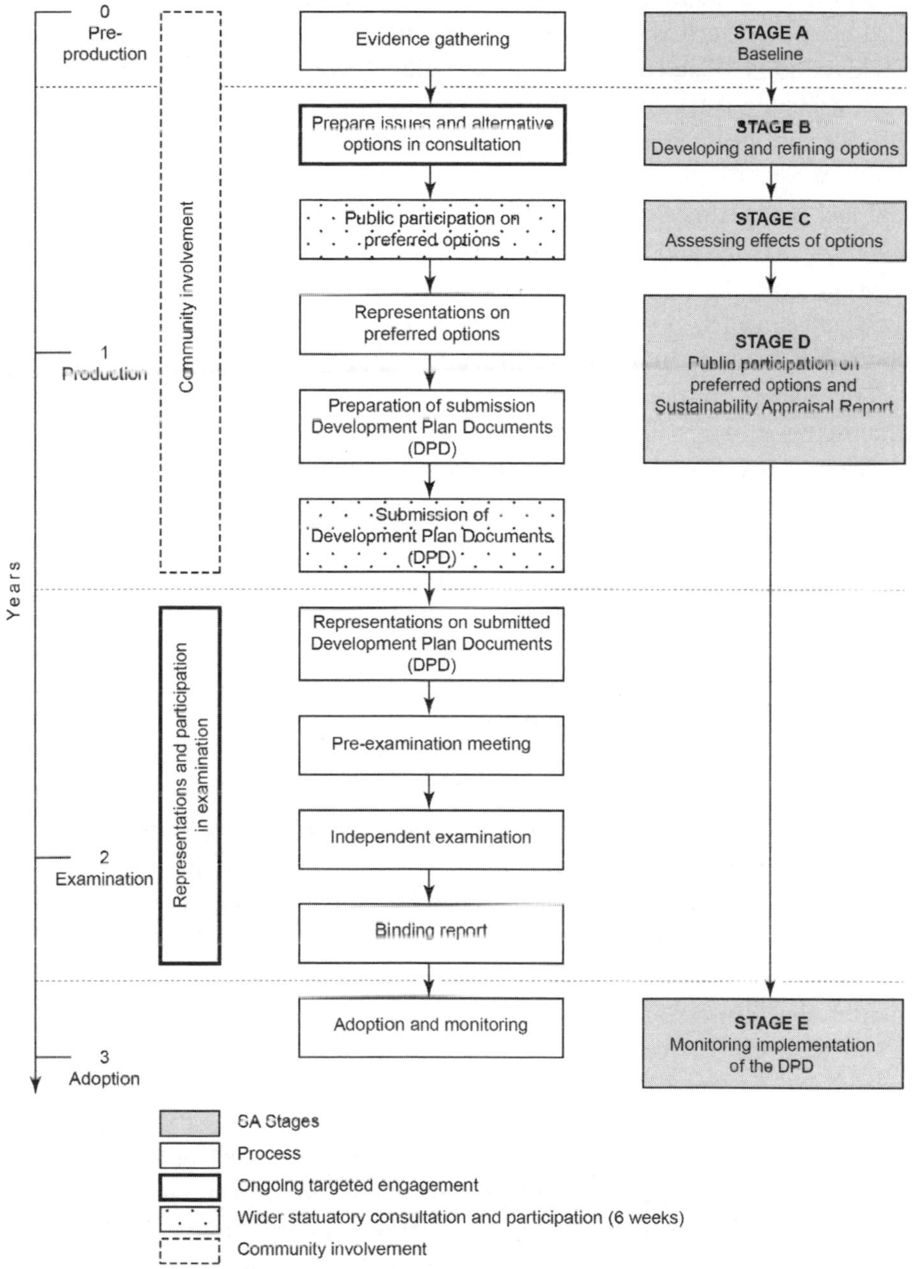

This emphasis on continuous monitoring reflects a frequent criticism of terrestrial planning processes that insufficient regard is given to issues of implementation. Led by authors such as John Freidmann (1967) and developed by Pressman and Wildavsky (1973) and others, Implementation Theory highlights the importance of considering issues of plan delivery at the earliest stages of the plan-making process, as a plan without the ability to implement it is meaningless (Berman, 1980; Weale, 1992). In addition, this work is significant in drawing attention to the fact that planners are involved in more than just plan-making. Crucially, *planning* activity also entails the realization of planning ambitions through action programmes and strategies of various sorts, or more typically through the slow, incremental, but very important process of decision-making related to individual licensing or development control applications. Although the style of development plans does vary, in many instances it is likely that they will not in themselves give a definitive answer as to the appropriateness of individual development proposals, and value judgements will therefore have to be made based on case-specific information, including the outputs of any public consultation processes. This is why many terrestrial planning systems involve elected representatives in this type of decision-making.

Before leaving this discussion of the terrestrial planning process, it is important to draw out this issue of variability in planning practice further, as it is potentially of great significance to the development of marine planning and management. For while many of the ideas and debates referred to above have had a worldwide reach, their local interpretation has varied substantially. This variation has especially been a focus of attention within the EU, where differing approaches to terrestrial planning have been identified as possibly a distorting feature of the EU free market and a factor impacting upon ambitions to achieve balanced regional development across the EU area. What has emerged from various research projects is a picture of distinct 'families' of planning within the EU, reflecting varying legal and administrative traditions in member states. Figure 2.4 graphically illustrates this point.

So what can marine planning and management draw from these debates related to the terrestrial planning process? Again, three examples serve to illustrate the potential. The first relates to the issue of the nature of the marine planning process, and the balance between scientific, communicative/collaborative, and design/arts inputs. To date, it seems fair to say, that unlike planning for the land, planning for the sea is developing from largely scientific roots. As Table 2.4 shows, many of the marine planning initiatives that are currently in place across the world are led by research institutes or government departments with a strong environmental/ecological remit. It is therefore unsurprising that scientific approaches to planning are to the fore.

Figure 2.4 Legal and administrative families of Europe

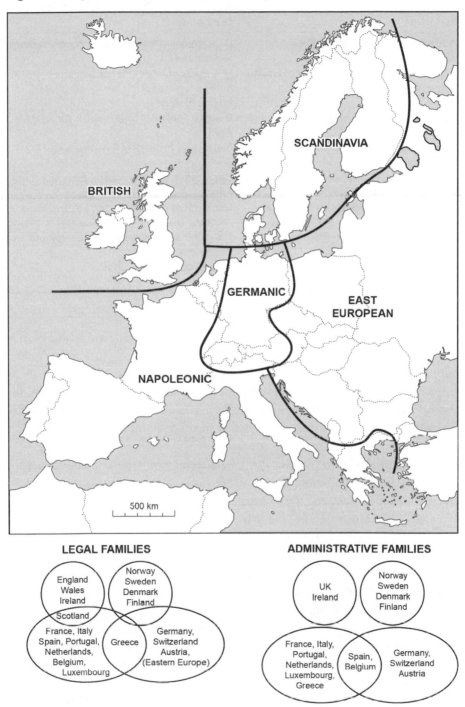

Source: Newman and Thornley (1996, p29)

Table 2.4 Some key marine initiatives and their lead bodies

Initiative	Lead body
Australia Marine Bioregional Plans	Department of Environment, Water Heritage and the Arts
A Spatial Structure Plan for the Belgium Part of the North Sea	University of Ghent Marine Institute and the Renard Centre for Marine Geology and Marine Biology
Spatial Planning for German North Sea and Baltic Sea	German Federal Maritime and Hydrographic Agency
Barents Sea-Lofoten Area Integrated Management Plan	Norwegian Ministry of Environment
UK Marine and Coastal Access Act	Department of Environment, Food and Rural Affairs
Large Ocean Management Area Integrated Management Plans	Department of Fisheries and Oceans, Canada
The Massachusetts Ocean Plan	Massachusetts Department of Energy and Environmental Affairs
Marine functional zoning in the Chinese Territorial Sea	State Oceanic Administration

Source: Derived from UNESCO (2010)

There is, however, certainly evidence of support for communicative/collaborative approaches in many instances. This reflects a widespread recognition that public support is important if marine plans and marine planning controls are to be implemented effectively. For example, one of the national guiding principles for coastal and marine spatial planning that have recently been issued by the White House Council on Environmental Quality in the US is as follows:

> *Coastal and Marine Spatial Planning development and implementation would ensure frequent and transparent broad-based, inclusive engagement of partners, the public, and stakeholders, including with those most impacted (or potentially impacted) by the planning process and with underserved communities. (White House Council on Environmental Quality, 2009, p7)*

This suggests that skills of facilitation, communication and mediation should feature within the skills set of marine spatial planners or perhaps form specialist inputs to marine planning teams.

There is less evidence at the present time that arts perspectives are also regarded as essential to the marine planning process, although there are examples that point the way to what contribution they could make. One of these relates to the work undertaken by the Countryside Council for Wales assessing the relative sensitivity of Welsh seascapes to offshore development which has drawn upon landscape assessment experience (CCW, 2010). As the scale of marine development grows, it is likely that it will be increasingly important that such place-making concerns are integrated into marine planning activity. An arts input of a different sort may also be helpful in facilitating meaningful stakeholder and community engagement with marine planning. In contrast to terrestrial planning situations, elected representatives, key stakeholders and the general public may be relatively uninformed about the nature of the marine environment and the planning and management issues to be addressed. Visualization and artistic interpretation techniques are beginning to be used to encourage public engagement in for example, planning for landscape change (Miller et al, 2010) and in other planning contexts and these have also been demonstrated to have a role marine contexts (North Kent Local Authority Arts Partnership, 2010). Beyond this, there is a growing recognition that analysis of historical texts may yield useful insights into the past ecology of marine areas and could be a useful source of qualitative data for marine spatial planning (for example, Robinson and Frid, 2008), so even historians may have a place within marine planning teams. Ongoing discussion regarding the appropriate mix of scientific, communicative and artistic skills required for effective marine planning and management therefore seems to be merited.

Secondly, it is evident from the brief discussion of debates related to the terrestrial planning process provided above, that matters of implementation are critical and that proper attention needs be given to these in developing marine planning and management arrangements. However, it could be argued that Freidmann's analysis of terrestrial planning in the late 1960s might apply to the marine context today. While much attention is being given to the process by which marine plans might be prepared, particularly upon ensuring their scientific rigour and public legitimacy, it seems that less attention has been paid to the realities of their future implementation. There is therefore often a gap between the content of marine plans and the ability to practically influence human use of marine areas in the prescribed way (Hinds, 2003). More than this, there seems to be an underlying assumption that the key activity for marine planners is making plans. Clearly this is an important task and given the absence of plans in many areas it certainly deserves some priority. However, it would be wrong to assume that once plans are in place that the main planning task is over, or that subsequent regulation of development will be straightforward. Experience from terrestrial planning indicates that planning agencies often need to take a lead during implementation, through proactive programmes of direct action to steer development in a desired direction and further thought would seem to be helpful regarding what these might entail and how they may be resourced. Equally, the activities of marine licensing, development control, or development management as it is sometimes known, and of enforcement should not be regarded as secondary to plan-making. Discussion regarding the requirements of these activities and the processes involved deserves just as much attention, as these are at the 'coal face' of marine planning where the ideas set out in plans are put into effect. In essence, the processes of planning need to be closely coupled institutionally with the arrangements for policy implementation, development regulation, monitoring and evaluation to create an integrated governance regime for the marine environment.

Thirdly, it seems that further exploration of the legal and administrative traditions of different marine regions will be important, so that new marine planning arrangements are carefully tailored to their local context. The great variation in terrestrial planning families even within an area as small as the EU indicates that proposals for generic marine planning arrangements should be viewed with caution. Instead, great sensitivity to the existing pattern of responsibilities in both marine and terrestrial areas seems to be called for, particularly where marine planning requires synergy that crosses national boundaries. This synergy, uniting different states in their efforts to achieve a common outcome, is likely to be encouraged in the European context by legislation such as the Water Framework Directive, and the Marine Strategy Framework Directive. These require consistent goals to be achieved in the marine environment, using common standards and approaches. Other parts of the world may have to rely instead on the more subtle overarching driving forces of the CBD and United Nations Convention on the Law of the Sea (UNCLOS). The following discussion of the current development of marine planning arrangements in the UK serves to illustrate many of the points set out above.

Origins of marine spatial planning in the UK

The approach to marine planning in the UK can be said to have been influenced by four sources of experience: UK terrestrial planning; international marine planning practice; existing sectoral approaches to planning of marine activities; and international and national marine policy development.

As indicated earlier in this chapter, terrestrial planning in the UK can trace its legislative roots to the early 20th century and is well-established as a mechanism which seeks to ensure that individual development proposals (incorporating changes of use as well as property development) are assessed for their social, economic and environmental impact. It has also become an accepted responsibility for communities to exercise decision-making powers over such proposals through elected local government. The context for decision-making is provided by comprehensive development plans, created locally but meeting national targets and criteria as expressed in national and regional guidance. To ensure continuity and justice an appeal system, operated on behalf of central government, supports the planning process allowing individual citizens and landowners to challenge the decisions made in an open way. In devising a marine planning regime for the UK, there has therefore been a wealth of experience of terrestrial plan-making and planning decision taking to draw upon. This, however, has required considerable adaptation to the marine context.

International marine planning experience is varied and culturally specific, responding to political, social and historic factors unique to the country concerned. The UK's Marine Spatial Planning – Irish Sea Pilot (MSPP Consortium, 2006) did, however, judge that a review of established marine planning activity would be useful and looked in particular at experience from Australia, North America, New Zealand, Europe, Fiji and the Philippines. Its findings indicated that many of the initiatives were still in the early phases of development and the practical issues associated with implementation were in many cases still being worked through. This has also been the conclusion of other recent reviews (for example, GHK Consulting Ltd, 2004). Equally, the review revealed that activity primarily related to the establishment of marine protected areas and marine reserves. It therefore tended to place conservation objectives above use related objectives and focus upon identifying additional controls on human activities to support the achievement of conservation goals. In addition, it was found that there was very limited reference to any economic assessment of how the objectives might affect human activities either directly or indirectly. While the experience was judged to provide useful guidance in relation to process matters and the detailed management of protected areas, it was considered to be of less assistance in guiding the comprehensive marine planning approach that was envisaged for the highly urbanized and intensively used waters of the UK. It was therefore concluded that findings from the review should be treated with some caution and it should not be assumed that experiences elsewhere could be directly transferable to the UK situation.

The UK approach to marine planning has always been intended to be comprehensive, covering all territorial waters and dealing with the management of future economic uses as much as with the conservation of environmental and historic resources. In this respect the UK's approach is very ambitious in an international context. It must also deal with a range of issues related to the different responsibilities of the devolved administrations of Wales, Scotland and Northern Ireland and reserved powers of central government, which together make for an administratively complex situation (see Figure 2.5).

In addition, the new system has had to be built around established sectoral marine planning arrangements. Different economic and environmental sectors of sea use and governance have, over the years, developed very different approaches to planning for the future of UK marine space. This has resulted in a complex and confusing pattern of zones prioritized for particular uses. For

Figure 2.5 *Marine planning in UK waters*

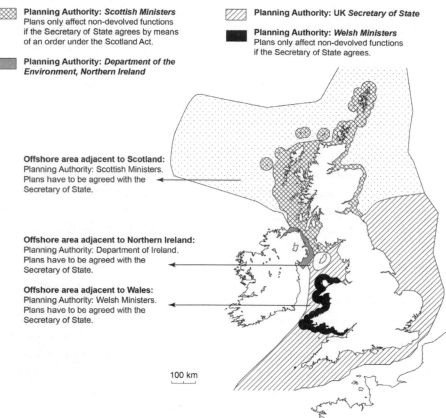

example, renewable energy has been a key area of attention and has involved the identification of offshore wind-farm areas of search by the Crown Estate. Other sectors where policy has been developed primarily in response to the needs of that sector are fisheries, marine aggregates, ports and shipping, oil and gas, and aquaculture. The requirement to undertake environmental assessments for proposals in each of these sectors illustrates how it is impossible to divorce one area of activity from others in decision-making, and also how much duplication is engendered by policy development in isolation; hence the realization that an integrated approach to planning is both more rational and also efficient. The Marine and Coastal Access Act, 2009, is a major step towards better integration. Although it carries forward some elements of sectorally based planning, including a requirement for the designation a network of Marine Conservation Zones based primarily on environmental criteria, it puts in place a new legal framework that should transform marine planning and management of UK seas very much in line with the ambitions of EA.

The Marine and Coastal Access Act reflects the growing recognition by the UK government over the last decade that there is now a need for a more integrated and sustainable approach to marine stewardship. This understanding has been underpinned by a succession of studies and reports from *Safeguarding Our Seas* (Defra, 2002) onwards. In addition, the UK has been responding to initiatives from the UN and EU in respect to marine stewardship. The OSPAR initiatives of the 1990s, for example, addressed risks related to single high profile thematic areas such as hazardous substances, eutrophication and biodiversity, while the Habitats and Birds Directives have led to the European network of marine protected areas, Natura 2000. Similarly the Common Fisheries Policy, Water Framework Directive, Renewables Directive, and Strategic Environmental Assessment Directive have all provided impetus to develop a marine planning system that integrates all competing interests and demands on a finite resource, the sea. More specifically OSPAR and the 5th North Sea Conference committed the UK and others to the concept of marine spatial planning and the EU Marine Strategy Framework Directive (MSFD), adopted in 2008, now requires that member states should achieve 'good environmental status' for their waters. The MSFD makes it clear that one of the tools by which member states can achieve Good Ecological Status (GES) is through marine planning. It is perhaps particularly important to note that the MSFD requires GES to be assessed at the level of a regional sea (rather than a member state's national waters), so implicit in this is an expectation that a certain level of trans-boundary cooperation and mutual support will become increasingly important.

The UK Marine and Coastal Access Act (2009)

This UK Marine and Coastal Access Act had its origins in a sequence of reports starting with the Marine Stewardship Report (Defra, 2002). A particularly significant contribution to the development of marine planning in the UK was also made by the Marine Spatial Planning – Irish Sea Pilot (MSPP Consortium, 2006). In addition to a review of international experience, the study simulated the production of a marine spatial plan for part of the Irish Sea (both plan production and stakeholder involvement) and provided a template regional sea level planning document which has been drawn upon by others in the UK and subsequently in Europe.

Following the publication of a consultation document in 2006 and the Marine Bill White Paper (Defra, 2007) the Marine and Coastal Access Act finally received Royal Assent in November 2009. The ingredients of the act are consistent with the earlier documents with the addition of the proposals to introduce a national coastal access corridor around the UK's coastline.

The key elements of the act are as follows:

- Establishing the Marine Management Organization (MMO);
- Establishing an Exclusive Economic Zone around the UK;
- Setting out a process for marine planning;
- Revision to the marine licensing regime;
- Establishing the process for designating Marine Conservation Zones;
- Introducing a series of measures in relation to fisheries including the replacement of Sea Fisheries Committees by Inshore Fisheries and Conservation Authorities;
- Modernizing the powers and processes of enforcement; and
- Improving access to the UK coastline for the public.

For the purposes of this chapter, the focus will be on marine planning but its operation cannot be divorced from other aspects of the act, particularly licensing, conservation, enforcement, coastal access and the key role of the MMO in undertaking the task of producing plans, licensing and enforcement.

The introduction of marine planning to UK waters is designed to apply the benefits of forward planning to the management of marine resources for both economic and conservation purposes. The sectoral approaches of the past have failed to provide sufficient strategic management of the UK's marine environment which, in line with the highly urbanized character of the country, have increasingly suffered environmental degradation in light of intense human pressure. The Marine and Coastal Access Bill – Policy Paper (Defra, 2009) sums it up as follows:

> *Marine Planning is essentially a process that will help us to be proactive about the way we use and protect our marine resources, and the interactions between different activities which affect them … It will create a framework for consistent and evidence-based decision-making, and through extensive public involvement will afford anyone with an interest in our seas the chance to shape how their marine environment is managed.*

The act establishes a requirement for a UK-wide marine policy statement (MPS). This statement will provide a basis for marine planning policies, how these will contribute to the achievement of sustainable development and setting out priorities that will then be detailed in a series of marine plans to be produced by plan authorities responsible for different regions of the UK waters. Essentially these authorities will be the MMO and the devolved administrations in Wales, Scotland and Northern Ireland. Given that the various devolved plan authorities have different intentions as regards their approach to plan-making, the negotiation of a consensual MPS represents a challenge.

The plan authorities will identify the appropriate areas to be covered by individual plans and this will be subject to consultation (Defra, 2009). Marine plans off the English coast will be produced incrementally although elsewhere, Scotland, for example, a broad brush strategic plan for all plan areas may be produced first.

The preparation of marine plans has been subject to early consideration and the outcome (see Figure 2.6) owes a great deal to terrestrial experience (see Figure 2.3 for a useful comparison) and the report of the Marine Spatial Planning – Irish Sea Pilot (MSPP Consortium, 2006). Particular features that are common to both terrestrial and marine planning in the UK include the requirement to undertake sustainability appraisal in parallel with plan preparation and also stakeholder and community involvement. To this end, a requirement of the Marine and Coastal Access Act is that a Statement of Public Participation should be published at the beginning of each plan-making process and there is advice on how best to achieve ongoing involvement through the use of advisory groups. Full consultation on draft plans must be undertaken before adoption and this process may also include independent scrutiny of the proposals. Subsequent

Figure 2.6 *The process for preparing UK marine plans*

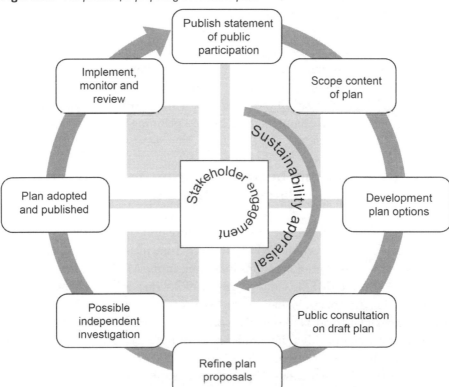

monitoring and review is also required of the plan-making authorities with the opportunity for further review, amendment or replacement over time.

Marine planning is, however, not just about making plans, and the day-to-day task of determining licences under the Marine and Coastal Access Act will require the MMO and others to make decisions in accordance with the MPS and marine plans. Licences previously granted under separate regimes will be consolidated in a single marine consent and consequently decisions on marine developments such as wind farms, tidal and wave power projects, jetties, moorings, aggregate extraction and dredging will be the responsibility of the MMO. Where the power to consent to new development lies outside the MMO's remit, such as oil and gas installations, large 'nationally significant' infrastructure projects, shipping and land-based activities with marine implications, those bodies will be required to have regard to the MPS and marine plans. Where necessary the MMO has enforcement powers to ensure compliance with its decisions.

Under the Marine and Coastal Access Act, the selection of sites for designation as marine conservation zones (MCZ) to protect rare, threatened and representative habitats and species, will be undertaken by the Secretary of State, Welsh and Scottish ministers acting on advice from statutory nature conservation bodies, such as Natural England, rather than by the MMO. Designations are due to be made by the end of 2012 and programmes are currently in place to identify potential sites through regional stakeholder projects. This means that designation will take place before marine plans are adopted and future plan-making and licensing will have to take account of these designations in the same way that they will have to retrofit to other designations such as the Round Three Off-shore Wind Licensing areas. In these respects, it will clearly take time for a fully integrated coherent and comprehensive marine planning system to emerge.

The relationship between marine and terrestrial planning in the UK

A particular issue where further thought and development is still needed relates to the relationship between the terrestrial and marine planning systems. This has been raised on a number of occasions including in the Marine Spatial Planning – Irish Sea Project (MSPP Consortium, 2006). That study proposed that a hierarchy of marine plans, including non-statutory documents such as Shoreline Management Plans and Harbour Plans should be established (see Figure 2.7) that would match up with a similar hierarchy of terrestrial plans. At the time of writing, in the English system, this matching hierarchy would include the Regional Spatial Strategy, Local Development Framework documents and any non-statutory plans including those that are coastal in nature.

Figure 2.7 *Suggested UK marine planning scheme*

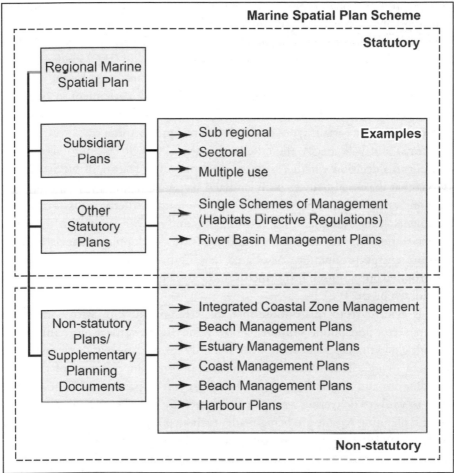

Source: MSPP Consortium (2006)

Integrated Coastal Zone Management (ICZM) strategies have also frequently been cited as offering an appropriate mechanism for coordinating plans either side of the coastal divide and steps have been taken towards this in the UK through the publication in 2009 of a Strategy for Promoting an Integrated Approach to the Management of Coastal Resources in England (Defra, 2009). Although this doesn't as yet closely specify how the interface between terrestrial, marine and costal planning regimes will be dealt with, it establishes an important mechanism through which this can be progressed.

There is, however, one level where recent legislation has established common planning arrangements for the land and the sea. The Planning Act (2008) introduces a new regime for nationally significant infrastructure project (NSIP) planning in England and Wales. Under the act, applications for the development of large projects (above a threshold size) for energy generation (including nuclear and renewables), pipelines, gas storage, electricity transmission, ports, strategic road and rail, airports, water supply, waste water and hazardous waste will be determined by a new Infrastructure Planning Commission (IPC) guided by new national policy statements (NPSs). The NPSs will be produced by the relevant government departments and subject to public and parliamentary scrutiny before being designated (adopted). The NPSs will provide both a justification of national need for certain types of development (and in some cases will identify locations) and will specify the criteria that the IPC will take into account in coming to a decision whether to grant consent. A key feature of NPSs and the IPC is that their remit covers both land and sea, most obviously in dealing with offshore renewables and ports but also where new infrastructure crosses the land/sea divide, as for example in the case of jetties and quays associated with coastal power stations or the connections and landside development associated with offshore energy production. Associated development (linked and necessary to the proposal but not itself infrastructure) will also be included in the applications considered by the IPC.

It is evident from this outline that the UK government, with widespread private sector and public support, is slowly moving towards a more integrated planning and management framework not only for the sea, but also crossing the land/sea divide. The difficulties in making progress in this area of complex jurisdictions and responsibilities are, however, very apparent. While important steps towards to delivering a simpler and hopefully stronger and more coherent marine planning regime are being made, each step along the way is raising new legal, administrative and political issues and complexities that need to be addressed. Linkage to the terrestrial planning system has certainly been beneficial in helping to guide the way forward. However, it is likely to be some time before the new marine planning system finally overcomes the legacy of sectoral planning and achieves coherence with planning for the land. Equally, while it finds its feet, the embryonic system is faced with a rapidly changing governance context which is requiring elements of rethinking and adjustment even in its hour of birth. A key development here has been the creation of the new Conservative Liberal coalition government in May 2010. This is bringing the prospect of major changes to the pattern of terrestrial planning including the abolition of the very recently created IPC and the regional tier of plan-making. This illustrates well the need for marine planning and management arrangements wherever they may be, to be closely tuned to the shifting spirit and architecture of governance. At the same time it also highlights the need to make progress towards better planning of our seas in less than perfect circumstances. As EA suggests, an adaptive management approach in the face of ongoing change and uncertainty is important and the UK experience reveals that this relates to both the natural and human aspects of marine ecosystems.

Conclusion

In conclusion, this chapter has sought to illustrate why close connection to spatial planning for the land may be of value in developing the human dimension of marine planning and management that is a key feature of EA. By providing a selected potted history of theoretical debates regarding the purpose and process of terrestrial planning, it has illustrated how this might be used to stimulate critical reflection upon the purpose and process of planning for the sea. It is encouraging that there is already evidence that, particularly in relation to matter of process, marine planning and management is in fact now beginning to draw quite heavily upon terrestrial planning practice and in particular on the current spatial planning paradigm. There is, however, perhaps a danger of adopting this paradigm in an unquestioning way without due recognition of key differences between terrestrial and marine environments, the history behind the development of spatial planning and the sophistication of the concept that draws together strands from a very long history of planning thought.

In terms of elaborating on the human dimension of EA with the benefit of marine planning in mind, this review of terrestrial planning has hopefully developed a deeper appreciation of how views on the purpose and process of any planning activity are likely to evolve over time which serves to illustrate the significance of Principle 1. This indicates that the objectives of planning and management of the sea are not value-free but are in fact matters of societal choice. These therefore require active debate and deliberation in order to achieve clarity and some degree of public consensus about the desired attributes to be strived for. In turn, reflections on terrestrial planning experience underline the importance of Principle 12 concerning the engagement of all sections of society (including the scientific community) in this process and how this should inform the skill set available within marine planning and management teams and the process through which decisions are made. Similarly, it reveals the relevance of Principle 11 that encourages the consideration of all forms of relevant information, including scientific and indigenous and local knowledge in decision-making. Perhaps most significantly though it draws attention to Principle 9 that encourages recognition that change is inevitable and indicates that this relates as much to the human as well as the natural dimensions of marine ecosystems. Interestingly here it highlights the need to look beyond charting changing patterns of human use of the sea, to the underlying cultural, legal, administrative, social and economic norms and practices that determine them. In this context, marine planning activity would do well to anticipate the shifts that may take place and, in line with Principle 6, set plans and objectives for the long term. It is perhaps critically important here to recognize that marine spatial planning is not simply about identifying and describing the factors of change and using them as inputs to decision-making. It is also worth remembering the value of the creative dimension of spatial planning which was historically so prominent and should not be forgotten or disparaged. At no time in human history has the need for creative thinking been more important. With the global population growing from 3 billion in the 1970s, to 6 billion at the present time, and with the prospect of 9 billion people by 2050, both sound science and imagination, as well as full engagement with the human dimension of EA will be needed in order to deliver more sustainable patterns of development in our seas.

References

Alexander, C. (1965) 'A city is not tree', *Architectural Forum*, vol 122, no 1. pp58–61 and vol 122, no 2, pp58–62

Berman, P. (1980) 'Thinking about programmed and adaptive implementation', in H. Ingram and D. Mann (eds) *Why Policies Succeed or Fail*, Sage, USA

Briassoulis, H. (1989) 'Theoretical orientations in environmental planning: An inquiry into alternative approaches', *Environmental Management*, vol 3, no 4, pp 381–392

CCW (Countryside Council for Wales) (2010) *Seascape Assessment of Wales*, CCW, Bangor

CEC (Commission of the European Communities) (2008a) *Towards a Future Maritime Policy for the Union: A European Vision for the Oceans and Seas*, Office for Official Publications of the European Communities, Luxembourg

CEC (Commission of the European Communities) (2008b) *Directive 2008/56/EC of the European Parliament and of the Council of 17 June 2008 establishing a framework for community action in the field of marine environmental policy* (Marine Strategy Framework Directive), Office for Official Publications of the European Communities, Luxembourg

Christensen, K. S. (1985) 'Coping with uncertainty in planning', *Journal of the American Planning Association*, vol 51, no 1, pp63–73

Committee of Ministers to Member States on the European Regional/Spatial Planning Charter (1983) *Spatial Planning Charter*, Council of Europe DG1V, Strasbourg

Cowell, R. and Martin, S. (2003) 'The joy of joining up: Modes of integrating the local government modernisation agenda', *Environment and Planning C: Government*, vol 21, no 1, pp159–179

Cullingworth, B. and Nadin, V. (2006) *Town and Country Planning in the UK*, Routledge, London

Defra (Department for Environment, Food and Rural Affairs) (2002) *Safeguarding Our Seas: A Strategy for the Conservation and Sustainable Development of our Marine Environment*, DEFRA, London

Defra (2007) *A Sea Change: A Marine Bill White Paper*, Defra, London

Defra (2009) *Our Seas – A Shared Resource: High Level Marine Objectives*, Defra, London

Douvere, F. (2008) 'The importance of marine spatial planning in advancing ecosystem-based sea use management', *Marine Policy*, vol 32, pp762–771

Dryzek, J. S. (1997) *The Politics of the Earth: Environmental Discourses*, Oxford University Press, Oxford

Ehler, C. and Douvere, F. (2007) *Visions for a Sea Change: Report of the First International Workshop on Marine Spatial Planning*, Intergovernmental Oceanographic Commission (IOC) and the Man and the Biosphere Programme, IOC Manual and Guides, No 48, IOCAM Dossier No 4, UNESCO, Paris

Etzioni, A. (1967) 'Mixed scanning: A third approach to decision making', *Public Administration Review*, vol 27, pp387–392

Freidmann, J. (1967) 'A conceptual model for the analysis of planning behaviour', *Administrative Science Quarterly*, vol 12, no 2, pp225–252

Frey, H. and Yaneske, P. (2007) *Visions of Sustainability: Cities and Regions*, Routledge, Abingdon

Gallent, N., Juntti, M., Kidd, S. and Shaw, D. (2008) *Introduction to Rural Planning*, Routledge, London

GHK Consulting Ltd. (2004) *Potential Benefits of Marine Spatial Planning to Economic Activity in the UK*, RSPB, Sandy, UK

Habermas, J. (1984) *Theory of Communicative Action*, Beacon Press, London

Hall, P. (2002) *Cities of Tomorrow: An Intellectual History of Urban Planning and Design in the Twentieth Century*, Blackwell, Oxford

Healey, P. (1997) *Collaborative Planning: Shaping Places in Fragmented Societies*, Macmillan Press, London

Healey, P. (2004) 'The treatment of space and place in the new strategic spatial planning in Europe', *International Journal of Urban and Regional Research*, vol 28, no 1, pp45–67

Hinds, L. (2003) 'Oceans governance and the implementation gap', *Marine Policy*, vol 27, no 4, pp349–356

Ingram, G. K., Carbonell, A., Hong, Y. H. and Flint, A. (eds) (2009) *Smart Growth Policies: An Evaluation of Programs and Outcomes*, Lincoln Institute of Land Policy, Cambridge, MA

Innes, J. E. and Booher, D. E. (1999) 'Consensus building and complex adaptive systems: A framework for evaluating collaborative planning', *Journal of the American Planning Association*, vol 65, no 4, pp412–423

ISOCARP (International Society of City and Regional Planners) (2009) *Low Carbon Cities: Review 05*, ISOCARP, The Hague

Jacobs, J. (1961) *The Death and Life of Great American Cities*, Random House, New York

Katz, P. (1994) *The New Urbanism: Toward an Architecture of Community*, McGraw-Hill, New York

Kidd, S. (2007) 'Towards a framework of integration in spatial planning: An exploration from a health perspective', *Planning Theory & Practice*, vol 8, no 2, pp161–181

Lindblom, C. (1959) 'The science of muddling through', *Public Administration Review*, vol 19, pp79–88

Luccarelli, M. (1995) *Lewis Mumford and the Ecological Region*, Guildford Press, New York

McLoughlin, B. (1969) *Urban and Regional Planning: A Systems Approach*, Faber & Faber, London

Miller, D., Fry, G., Quine, C. and Morrice, J. (2010) *Managing and Planning Landscape Change: The Role of Visualisation Tools for Public Participation*, Springer, New York

MSPP Consortium (2006) *Marine Spatial Planning Pilot*, MSPP, London

Mumford, L. (1961) *The City in History: Its Origins, its Transformations, and its Prospects*, Penguin, Harmondsworth

Neal, P. (2003) *Urban Villages and the Making of Communities*, Spon, London

Newman, P. and Kenworthy, J. R. (1999) *Sustainability and Cities: Overcoming Automobile Dependence*, Island, Washington, DC

Newman, P. and Thornley, A. (1996) *Urban Planning in Europe: International Competition, National Systems and Planning*, Routledge, London

North Kent Local Authority Arts Partnership (2010) *Vanishing Shores*, www.nklaap.com/vanishingShores.html, accessed 15 May 2010

ODPM (Office of the Deputy Prime Minister) (2005) *Local Development Framework Monitoring: Good Practice Guide*, ODPM, London

Pressman, J. L. and Wildavsky, A. (1973) *Implementation: How Great Expectations in Washington Are Dashed in Oakland*, University of California Press, California

Robinson, L. and Frid, C. (2008) 'Historical marine ecology: Examining the role of fisheries in changes in North Sea benthos', *Ambio: A Journal of the Human Environment*, vol 37, no 5, pp362–372

Schön, P. (2005) 'Territorial cohesion in Europe?', *Planning Theory & Practice*, vol 6, no 3, pp389–400

Taylor, N. (1998) *Urban Planning Theory Since 1945*, Sage Publications, London

Tewdwr-Jones, M. and Williams, R. H. (2001) *The European Dimension of British Planning*, Spon, London

Trist, E. (1980) 'The environment and system-response capability', *Futures*, vol 12, no 2, pp113–127

UN (United Nations) (1982) *United Nations Convention on the Law of the Sea*, UN, New York

UNEP (United Nations Environment Programme) (2006) *Marine and Coastal Ecosystems and Human Well-being: A Synthesis Report Based on the Findings of the Millennium Ecosystems Assessment*, UNEP, Nairobi

UNESCO (United Nations Educational, Scientific and Cultural Organization) (2010) *Marine Spatial Planning*, www.ioc-unesco.org/index.php?option=com_frontpage&Itemid=1, accessed 15 May 2010

Watson, N. (2004) 'Integrated river basin management: A case for collaboration', *International Journal of River Basin Management*, vol 2, no 3, pp1–15

Watson, N., Deeming, H. and Treffny, R. (2009) 'Beyond bureaucracy? Assessing institutional change in the governance of water in England', *Water Alternatives*, vol 2, no 3, pp448–460

Weale, A. (1992) 'Implementation: A suitable case for review?', in E. Lykke (ed) *Achieving Environmental Goals*, Belhaven, London

White House Council on Environmental Quality (2009) *Interim Framework for Effective Coastal and Marine Spatial Planning*, WCEQ, Washington, DC

World Commission on Environment and Development (1987) *Our Common Future*, Oxford University Press, Oxford

Chapter 3

EU Maritime Policy and Economic Development of the European Seas

Greg Lloyd, Hance D. Smith, Rhoda C. Ballinger, Tim A. Stojanovic and Robert Duck

This chapter aims to:

- Examine European Union (EU) maritime policy contexts, including the patterns of sea uses and regional development, and related environmental and institutional dimensions;
- Review the development of EU marine and maritime policy;
- Consider wider EU constitutional and political developments, with particular reference to the Lisbon Strategy;
- Analyse the changing contexts of ideology and political development; and
- Discuss interrelationships between EU development and maritime policy, including the ecosystem approach context, economic and social change, environmental management and planning, and maritime policy outcomes.

Introduction

The economic development of the European Economic Community (EEC) was established by the Treaty of Rome in 1957; its predecessors with their roots in the late 1940s and early 1950s – notably the Organisation for European Economic Co-operation in 1948 and the European Coal and Steel Community in 1952; and the subsequent European Union (EU) can usefully be considered in two long stages. The first of these, from the late 1940s until the late 1980s/ early 1990s is associated with the establishment of the Common Market, the elaboration of the major European institutions of state governance, and geographical expansion to encompass both the southern periphery (Greece in 1981, and Spain and Portugal in 1986) and the northern periphery of Sweden and Finland together with Austria in 1995. The second major phase – still in its early stages – was signalled by the collapse of the Berlin Wall in 1989 and the subsequent dissolution of the Soviet Union, paving the way for the integration of the vast eastern periphery in the first decade of the 21st century. Although the landmarks were political, the essence of the stages is economic – successive complex combinations of economic development associated with substantial social and political changes and accompanied by increasing environmental pressures.

It is in this broad context that the evolution of both EEC and EU maritime policy and its relationships with economic, political and legal evolution of the EU can be best understood. During the first stage there was no maritime policy as such, either at European level or at national levels. In both cases, policies and management relating to the sea were very largely sectoral, accompanied by regional international arrangements covering the North East Atlantic and its adjacent seas, notably in the fields of ports and shipping, military strategy, fisheries, marine environment, and marine science. By far the most significant EU maritime component was the Common Fisheries Policy (CFP), conceived in conjunction with the Common Agricultural Policy (CAP) during the first stage of development of the EEC between 1957 and the accession of the UK, Ireland and Denmark in 1973. It took ten years from this enlargement to renegotiate the CFP as the three nations possessed by far the greatest share of fisheries resources of the enlarged Community. During this time, the fisheries resources underwent continuous and significant decline. Most notably herring fishing in the North Sea was banned between 1977 and 1983 in the wake of excessive exploitation by the latest generation of purse seiners, while major demersal stocks underwent continuous decline.

The second stage of development is characterized by three primary themes in the present context. The first concerns the advent of major environmental legislation at both national and European levels, with the European initiatives increasingly setting the pace for good environmental management. Of particular note have been the Directives: Wild Birds, Habitats, Urban Wastewater and Bathing Water Quality; the Water Framework Directive; and the Marine Strategy Framework Directive. The second theme has been the push towards greater economic integration exemplified by both the Gothenburg and Lisbon Strategies, and paralleled by the institution-building enshrined in the Maastricht, Amsterdam, Nice and Lisbon treaties. These developments have been paralleled by greater devolution within a number of member states, notably the UK, France and Spain. The third theme has been the divergence of interests between the core and peripheral regions of Europe, which is central to the understanding of economic and political change, and particularly the drive to create a fully integrated European state spearheaded by the Franco–German axis, and resisted to varying degrees by the vast maritime periphery. Examples of this resistance include, for example, the obstacles to UK entry during the first stage of development of the EEC, the non-entry of Norway to the EEC and the special arrangements made for the Faroe Islands and Greenland to remain outside the EEC when Denmark entered.

The purpose of this chapter is accordingly to begin with an examination of the maritime policy contexts, including the patterns of sea uses and regional development, and related environmental and institutional dimensions. The main focus is then concerned respectively with EU marine and maritime policy development during the second phase of EU development. This is followed by consideration of wider EU constitutional and political developments, with particular reference to the Lisbon Strategy. There follows a detailed analysis of the changing contexts of ideology and political development. Finally a discussion of interrelationships between EU development and maritime policy, including the ecosystem approach (EA) context, economic and social change, environmental management and planning, and maritime policy outcomes.

Maritime policy contexts

Understanding the maritime policy contexts involves: interrelationships among sectorally based sea and coastal uses; regional economic and social development; environmental influences and impacts; and political, legal and institutional change, all of which constitute elements in marine policy, governance and management. The approach adopted in this section is to provide a regional overview of these interrelationships.

The starting point is consideration of regional patterns of sea use, based on the classification of fundamental groups presented in Figure 3.1 (see, for example, Smith, 1991). The transport and (military) strategic groups may be primarily defined in terms of spatial organization and incorporate the spatial organization components of the remaining groups. Of particular note are the locations of ports, all types of shipping routes, naval exercise areas and submarine cables. There is a strong correlation with overall geographical patterns of economic development discussed below. For mineral and energy, and living resources, the defining features of the system revolve around interactions between human activities and the marine environment stemming from the exploitation of the resources concerned for food, materials and energy. In contrast to the first group, defining features are the correlations with locational patterns of natural resources and other environmental features. For the remaining groups of uses common to sea and land the primary defining features in locational terms relate to human behaviour and decision-making, with complex patterns aligned both to economic development and environmental characteristics. The 'land only' categories referred to in Figure 3.1 are of primary interest in terms of coastal location.

Figure 3.1 *Sea and coastal uses and management*

			Sea and land									Land only	
			Transport	Strategic	Minerals & energy	Living resources	Waste disposal	Leisure & recreation	Education & research	Conservation	Coastal engineering	Settlement	Manufacturing & services
TECHNICAL MANAGEMENT	Information management	Environmental monitoring											
		Surveillance of uses											
		Information technology											
	Information assessment	Environment											
		Technology											
		Economic											
		Social											
		Risk											
	Professional practice	Natural/social sciences											
		Surveying											
		Engineering											
		Accountancy											
		Planning											
		Law											
GENERAL MANAGEMENT		Technical management coordination											
		Organization management											
		Policy											
		Strategic planning											

Note: The uses classification appears across the top of the table. The management classification extends from physical interactions between uses on the one hand and the environment on the other – the technical management dimension – to the human dimensions of general management, concerned with governance.

A useful reference for the matrix is Smith H. D. (1992) 'Theory of ocean management' in Paolo Fabbri (ed) *Ocean Management in Global Change*, London and New York, Elsevier Applied Science pp9–38.

The starting point for regional analysis of sea uses is an acknowledgement of the primacy of economic and social development as a determinant of regional patterns, temporally defined in terms of the two stages referred to in the Introduction above. Even a cursory investigation of, for example, the CFP underlines the primary influence of economic factors in both the development of the fishing industry and the failure of those aspects of the policy dealing with conservation of fish stocks. The regionalization of Europe adopted here is illustrated in Figure 3.2. There is a broad distinction between an economic core region on the one hand, and a vast, substantially maritime periphery on the other (Ballinger et al, 1994). The principal factors taken into account in this regionalization include: distribution of population and both coastal and inland settlement; economic activity, including land, air and maritime transport and ports; and manufacturing and leisure industries – especially those located at the coast. There is inevitably a degree of interpretation and fuzziness in the transition between the two regions, as illustrated in Figure 3.2.

In a maritime context, apart from the high densities of coastal settlement and coastal engineering development, the core is characterized by: possession of a majority of the largest ports, a number of which are connected to their hinterlands by large rivers and canals; major concentrations of coastal manufacturing and energy generation plants; offshore aggregate dredging; many distant-water fishing ports that developed during the first temporal stages noted above; and a high concentration of coastal leisure industries. All of these produce high levels of environmental impacts. By contrast, the maritime peripheral regions are associated with a predominance of resource extraction, including the offshore oil and gas industry, pelagic and demersal fisheries and fish farming. Substantial port and shipping industries and leisure activities are often more regionally concentrated. The maritime and coastal economies are frequently the dominant element, especially in the Scandinavian countries and parts of the Atlantic and Mediterranean, a fact acknowledged, for example, in the emerging EU Maritime Policy, the Common Fisheries Policy and the activities of the European Regional Development Fund (ERDF) such as the Interreg research programmes. The core–periphery use patterns are replicated to varying extents at subregional level, especially in the periphery, centred round large port cities and towns.

Not surprisingly the marine and coastal environment of Europe is complex at a number of geographical scales. A first order classification of coasts is illustrated in Figure 3.2 (Ballinger et al, 1994). In the north is the complex of fjord and fjard coasts of mainland Scandinavia, Finland, Iceland, Scotland and the northern half of Ireland, which are predominantly rocky, both sheltered and exposed to the open sea. The central division extends from the eastern English Channel through the southern North Sea to the southern Baltic, characterized by low sandy coasts and barrier islands exposed to the open sea and intercalated with large estuaries. The western coasts coincide with the Armorican rocks and include both open rocky coasts and sheltered rias. The southern coasts are the mountainous, partly tectonically active Mediterranean and Black Sea coasts, where open rocky coasts and beach development are characteristic.

Figure 3.2 European regional maritime development

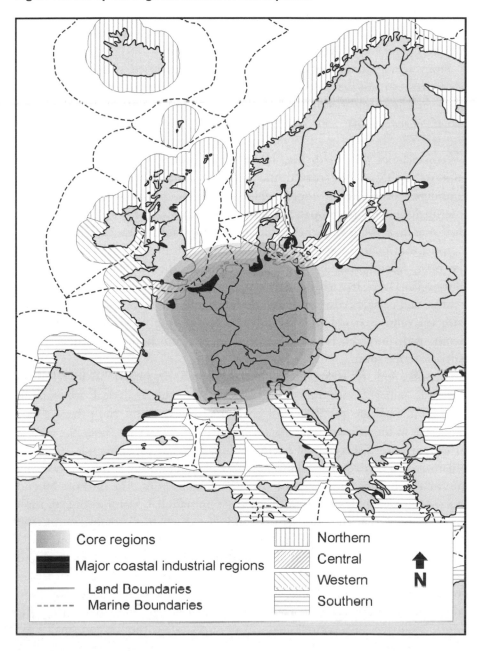

The coastal geomorphological divisions are reflected offshore, notably in the deep waters of the Norwegian Sea, Norwegian Deep and The Minches to the north; the steeply shelving coasts of Iceland, the Faroe Islands and submarine banks associated with the mid-Atlantic Ridge to the north-west and the similar topography of the Azores, Madeira and the Canaries to the south-west; the shallow waters of the southern North Sea, eastern English Channel and southern Baltic in the centre; the Celtic Sea, Bay of Biscay and mainland Portuguese coast to the west; and the narrow continental shelf and deep ocean basins of the Mediterranean and Black Sea.

From a biogeographical perspective the fully enclosed seas of the Baltic, Mediterranean and Black constitute large marine ecosystems in their own right. There are two truly oceanic large marine ecosystems with a transition zone roughly between 50° and 52° N in the open ocean to the west of Britain and Ireland (Sherman et al, 1993; see also Spalding et al, 2007). The northern system includes the North Sea. However, in development and management contexts, nearly all the coastal and marine environments are more usefully considered primarily in topographic and geomorphological terms and at intermediate scales encompassing individual estuaries, fjords and fjards; waters shallower than 100m with their associated fishing banks – especially in northern Europe, continental slopes and deep basins of both the fully enclosed seas and the open Atlantic and North Sea, for example, the Bay of Biscay and the Norwegian Deep. It is particularly notable that all of these have characteristic benthic and pelagic ecosystems; and that the core region coincides substantially with the central coastal environment and its offshore waters less than 100m depth, with only the Ligurian Sea and shallow northern extremities of the Adriatic to the south.

Identification of stakeholders and associated organizations involved in maritime policy is not inherently difficult, but the analytical approach is important. First are the drivers of economic development, the private sector industries in the primary, manufacturing and tertiary sectors. Since the demise of distant water fisheries, the fishing industry on the catching side is mainly organized as a myriad of small businesses, while fish farming and fish processing are the province of large, partly transnational companies. The offshore oil and marine aggregate industries have distinctive transnational corporate organization on a very large scale. In manufacturing and related activities, notable marine and marine-related industries include shipbuilding, coastal power generation and a range of heavy industries such as oil and petrochemicals, fertilizers, steel and other metal manufacturing. In the service sector, ports are characterized by a complicated mix of public and private ownership, shipping and telecommunications are large-scale transnational for the most part, while the leisure industries are a complicated mix of large-scale corporate and small-scale business organizations. The military and education sectors are mainly state-run, while conservation is a mix of state and voluntary sector organizations.

In the gradual transition from conventional state government to wider concepts and approaches to governance involving private, state and voluntary sector organizations as well as public participation, as far as the private sector is concerned the offshore oil industry in the North Sea has led the way with coordinated interaction both with other sea users and government, for example, through the UK Offshore Operators Association (UKOOA), now UK Oil & Gas. European ports have also taken an increasingly high profile role in environmental management (Ecoports, 2010). Shipowners tend to operate directly with government through consultation, participation and lobbying activities. The catching side of the fishing industry by contrast has hitherto relied to a substantial extent on lobbying, at both national and EU levels.

In the public sector, the principal stakeholders are agencies with specific sectoral responsibilities, especially for fisheries, environment and conservation together with marine spatial planning, with central government departments in direct control in all sectors for policy and administration. Local government has primary responsibilities for waste disposal, leisure and recreation, education, land-use planning and some coastal engineering. The EU's principal roles are focused on fisheries and environment, including conservation, with hitherto more limited roles in other areas such as maritime transport and ports.

Apart from the maritime safety role of the Royal National Lifeboat Institution (RNLI) in the UK and Ireland, the voluntary sector organizations and the general public – civil society – roles are mainly focused on conservation, of both the natural and cultural heritage. In particular, there is a strong tradition of heritage conservation in various parts of Europe in relation to certain sectors, of which ports, shipping, navies and fisheries are the most important.

The EU role in maritime policy has only become evident very recently in the sense of the development of a coordinated overview and priority actions engendered both by the Maritime Policy itself and activities surrounding the development and implementation of the Strategic Marine Framework Directive. Far more important has been the development of the Common Fisheries Policy and a range of directives applicable to the coastal and marine environment, together with the Recommendation on Integrated Coastal Zone Management. It is to these overarching developments that we now turn.

EU marine and maritime policy development

There are three principal components of maritime governance that merit particular attention in the present context. The first and most extensive is policy, which includes policy objectives, legal provisions, and financial instruments, all managed by the EU centrally through its institutions, particularly the Council of Ministers and the Directorates-General (DGs). Policies involve substantial derogation of sovereignty by member states, who are compelled to enact policy provisions into national legislation. By far the most important marine component is CFP, elaborated below, and presently administered by the Directorate-General for Marine Affairs and Fisheries. The second element is the directives, or regulations approved by the European Parliament and Council of Ministers. A number of these have been enacted over a period of time that have important maritime implications, relating especially to the marine environment, culminating in the Marine Strategy Framework Directive. These also require adoption by member states' legislatures. The third element is the recommendations, which are not legally binding, but can be widely adopted by member states as contributing to best practice in management and administration in particular fields and, as such, can exert substantial political influence in the conduct of EU affairs at both member state and EU levels. The most important example in the present context is the Recommendation on Integrated Coastal Zone Management of 2002.

The CFP (Wise, 1984; Holden, 1994) had its origins in 1970 in the adoption by the then member states – only six – principally located in the core region, with substantial maritime interests which nonetheless were relatively limited compared to those of the present 27. Measures included agreement in principle that European Commission (EC) fishermen should have equal access to member states' waters subject to reservation of small-scale coastal fisheries' traditional grounds. Other measures included a common market for fisheries products, and a structural policy to support investment in fishing vessels and onshore installations – in other words, both the catching and shore side of the industrial infrastructure. The policy became much more important with the entry of the UK, Ireland and Denmark in 1973: the three new member states possessed around 60 per cent of the fisheries resources of the enlarged Community. This, coupled with the extension in 1976 of member states' fisheries jurisdiction from 12–200 nautical miles, led to long and difficult negotiations leading to the introduction of the first comprehensive iteration of the CFP in 1983. It underwent some revision in 1992 and radical revision in 2002, and is due to complete another iteration by 2012 (EC, 2010). Here, discussion remains focused on fisheries management offshore and, to a limited extent, related onshore implications with regard to the CFP currently in force.

Despite the advent of increasingly sophisticated iterations of the policy, the great majority of commercial fish stocks have been subject to decline and instability from the beginning. Examples include the severe pressure exerted on North Sea herring fisheries by purse seining from the late 1960s onwards, which led to banning of herring fishing in the North Sea between 1977 and 1983, followed by continuous fishing pressure on the stocks ever since. Most major demersal stocks have also been subject to more or less continuous decline. By the time of the Green Paper on the CFP, published in 2009, some 88 per cent of Community stocks had been fished beyond the maximum sustainable yield (MSY), with 30 per cent of these stocks beyond safe biological limits, although there are substantial regional variations related to the complex geography outlined above.

The current policy aims to move towards longer term management using recovery and management plans, integration of environmental considerations into fisheries management, establishment of Regional Advisory Councils based on physical sea regions to increase stakeholder participation, continuing to promote fleet reduction, and limitation of fishing effort using technical management measures such as days-at-sea provisions used in the context of multi-annual guidance plans. Despite continuation of a high level of state support paid in aid and subsidies, and also including payment for fisheries management overall, the Commission has identified structural weaknesses to be addressed by 2012 including the continued problem of fleet overcapacity, imprecise policy objectives, short-term decision-making, insufficient responsibility given to the industry itself for management decision-making, and a lack of political will coupled with poor industry compliance (EC, 2008). Among the measures likely to emerge for the next stage of development of the policy are the introduction of Individual Transferable Rights (ITRs) with regard to fishing quotas, a much enhanced role for fishermen in decision-making and a substantial strengthening of regional management arrangements.

It would be difficult to exaggerate the seriousness of these circumstances. Fisheries lie at the very heart of the impact of human activity on marine ecosystems in the European region, and have undoubtedly impacted these ecosystems to a dangerous level of instability in many cases. And yet fisheries are also pivotal to a very large number of coastal communities in the vast maritime periphery of Europe, not least the largest fishing nations of Norway and Spain. Within the EU, this is already acknowledged in the instigation of the FARNET system by the Commission, concerned with promotion of alternative economic activities within fishing regions (EC, 2010).

In addition to the EU's Common Fisheries Policy, which dates back to 1970, there has been a burgeoning of European legislation and policy of relevance to the management of coastal and marine areas over the last few decades. This includes policies such as the European Spatial Development Perspective and the Sustainable Development Strategy which promote balanced and sustainable socio-economic development. The plethora of legal instruments contain Directives, requiring national implementing measures, and Regulations, which operate within existing laws, in addition to 'soft', less formal Recommendations, which promote new practices and approaches. As noted by Long and O'Hagan (2005), this complex and sectorally based legal framework and its associated law-making processes pose particular issues for the development of integrated marine management and planning. However, while there are complexities and difficulties associated with the decentralization of European policy through the transposition of directives into national legislation (Brinkhorst, 1999; Haigh, 1999), the European dimension still remains a key driver for improved environmental management and planning at national and sub-national levels, as exemplified by the recent paper concerning the Severn Estuary by Ballinger and Stojanovic (2010).

While early European environmental legislation of the 1970s and early 1980s was confined to a few discrete areas of pollution control and waste management, focusing on discharges from point sources and public health protection, increasing provisions over subsequent decades resulted in a tranche of directives of increasing stringency and scope (Jordan, 1998). Subsequent 'second generation' policy development saw a gradual shift towards addressing more complex environmental matters, addressing diffuse pollution, integrated pollution control and process standards (Bell and McGillivray, 2006) as well as the protection of ecological communities through the Habitats Directive (92/43/EEC) and the Convention on Biological Diversity (1992) (Elliott et al, 1999). This 'second generation' also saw the introduction of cross-sectoral environmental measures including the Environmental Impact (Directive 86/337/EEC and Amendment 97/11) and Strategic Environmental Impact (Directive 2001/42/EC) directives.

Post-1990, the scope of environmental legislation widened still further alongside legislative refinement and regulatory consolidation (Jordan, 1998), typified by the substantive and innovative Water Framework Directive (2000/60) (Page and Kaika, 2003), alongside the gradual embedding of environmental provisions within sectoral European agricultural, fisheries and shipping policy (Ballinger and Stojanovic, 2010). New regulations, standards and priorities marked significant change to the style, process and aspirations of environmental and sectoral management regimes, reflecting the new underlying paradigms associated with the concepts of sustainable development and the ecosystem approach. These encouraged a more holistic, system-based approach to planning and management (Ballinger and Stojanovic, 2010). This was further supported by the Treaty of Amsterdam 1999, requiring environmental protection to be integrated into all Community policies and activities (O'Hagan and Ballinger, 2009). However, recent evaluation of the implementation of European legislation, such as that completed within the Interreg IIIb COREPOINT project, suggest that the large and complex European legislative and policy framework poses significant constraints on local integrated coastal management and planning (O'Hagan et al, 2005).

Within the last decade new and emerging policy areas, notably those related to flood hazards, climate change, spatial planning, Integrated Coastal Zone Management (ICZM), the marine environment and maritime affairs have been added to the European policy framework. Many of these challenge traditional thinking and approaches to environmental management as well as sectoral management regimes, requiring closer integration and coherence between traditionally disparate policy areas. Of particular note in the context of marine spatial planning, are the Recommendation concerning the implementation of ICZM in Europe (2002/413/EC), the Marine Strategy Framework Directive (2008/56/EC) and the Integrated Maritime Policy of the European Union (COM (2007) 575 final). While not a legally binding instrument, the former remains influential for European coastal management, providing principles for good coastal management, and having been supported within the text of the European Spatial Development Perspective. The Integrated Maritime Policy highlights the business opportunities for different marine sectors and develops a governance framework through a variety of marine initiatives which are aimed at improving the efficiency and effectiveness of marine governance in Europe, including: a European Maritime Transport Space without barriers; a European Strategy for Marine Research; national integrated maritime policies to be developed by member states; a European network for maritime surveillance; a roadmap towards maritime spatial planning by member states; a strategy to mitigate the effects of climate change on coastal regions; reduction of carbon dioxide (CO_2) emissions and pollution by shipping; elimination of pirate fishing and destructive high seas bottom trawling; a European network of maritime clusters; and a review of EU labour law exemptions for the shipping and fishing sectors.

The Marine Strategy Framework Directive is significant in proposing an EA and environmental management of European seas on a regional basis. EA is held to represent a potential paradigm shift in the management of the natural environment and its constituent resources which are derived from the functioning of component ecosystems. This potential paradigm shift is based on a number of key premises which recognize that (i) sustainability of economic systems and quality of human life is inevitably dependent on the maintenance of healthy ecosystems; (ii) that humans are an integral part of ecosystems rather than separate from them; and (iii) a sectoral approach to management is generally insufficient to deal with the complex interrelationships and diverse stakeholder priorities that exist in the real world. This represents a relatively more holistic set of ideas, values and assumptions about the natural environment, the associated state–market–civil relations and the necessary institutional capacities and forms of intervention that are required to understand the value of the natural environment to society as a whole.

EU constitutional and political development and the Lisbon Agenda

The emergence of the second stage of development of the EU in the course of the 1990s has revolved around processes of enlargement through accession of new countries accompanied by a series of constitutional amendments partly designed to cope with the process of enlargement while at the same time providing increasing levels of integration of EU functions. The countries involved have already been referred to in the introduction to this chapter. The constitutional changes are enshrined in the successive treaties: Maastricht (1992), Amsterdam (1999), Nice (2003) and Lisbon (2009). In the decade to 2010, these developments have been supplemented by two strategies introduced by the European Commission: the Gothenburg Strategy dealing with economic, social and environmental sustainability; and the Lisbon Strategy or Agenda focusing particularly on innovation as a path to a more competitive and dynamic economy. The two strategies were intended to be complementary.

The Lisbon Agenda (2000–2010) was intended to provide a policy and institutional framework, principally concerned with stimulating economic growth and jobs. Its aim was to make the EU 'the most dynamic and competitive knowledge-based economy in the world capable of sustainable economic growth with more and better jobs and greater social cohesion, and respect for the environment by 2010', set against the background of productivity in the EU being below that of the US. It thus asserts a particular set of values, assumptions and priorities regarding the relationship between economic growth and development agendas and the natural terrestrial and marine environments. This reflects the market principle of economic competitiveness that permeates the design and thinking of the initiative. As such it was both an economic imperative and an economic narrative that set the context for other EU social and environmental policies. It was intended to deal with the low productivity and perceived stagnation of economic growth in the EU through the formulation of various policy initiatives to be adopted by all member states. This was a strategic 'call to arms' by which the economic framework for action was devised within which interventions by the individual nation states in the EU were then expected to comply. The broader objectives set out by the Lisbon Agenda were expected to be attained by 2010.

The main parameters of the Lisbon Agenda address economic, social and environmental renewal and sustainability considerations. In practice, however, the Lisbon Agenda is heavily based on the economic concepts of innovation as the perceived motor for promoting economic change and securing economic growth, the promotion of the 'learning economy' in terms of skills and adaptation to changing circumstances and opportunities, and enabling social and environmental renewal in those areas or sectors that were underperforming relative to the EU averages. The thinking behind the Lisbon Agenda was clear – it was to promote an economic imperative to achieve higher levels of growth and development, investment and productivity, and to support a general improvement in the economic viability and vitality of the EU as a whole. Importantly, however, under the Lisbon Agenda, it was anticipated that a stronger economy would create employment in the EU, alongside inclusive social and environmental policies, which would themselves drive economic growth even further.

In many ways the Lisbon Agenda represents a very conventional approach to spatial economic policy. It is principally 'top-down' in character, it seeks to devise a strategic framework for sub-EU and sub-national actions, it seeks to promote economic growth as a whole and then rely on 'trickle down' to address broader social and environmental issues. This was generally the case followed in individual EU nation states – such as the UK – in the earlier post-war period (Armstrong and Taylor, 2000). Importantly, commentators have pointed out that the basic assumptions of this particular form of market-influenced policy and institutional intervention was based on a low level of ecological consciousness and emphasized the exploitation and development of natural resources to secure the ambitions for economic growth (Giddens, 1998). This would suggest that history is simply perpetuating itself in terms of the dysfunctional relationship between economic and environmental agendas.

In 2004, the European Council and the Commission decided to prepare a mid-term review of the Lisbon process. This was to be presented to the Spring Summit in March 2005. Former Dutch Prime Minister Wim Kok was mandated by the March 2004 European Council reviewing the Lisbon Agenda and its progress. The Kok Report concluded in November 2004 that relatively little progress had been made over the first five years and recommended that the action agenda be refocused on securing relatively greater economic growth and employment. It asserted that progress under the Lisbon Agenda required active engagement by the member states of the economic and institutional reforms needed. One of the main conclusions of the Kok Report was that 'the promotion of growth and employment in Europe is the next great European project'. It argued that the economic progress achieved was hitherto 'unconvincing' and required reform. This reform followed when the European Parliament promoted actions to further stimulate growth and employment but with the caveat that such development should be fully consistent with the broader policy objective of sustainable development.

In its resolution on the mid-term review of the Lisbon Agenda in March 2005, the European Parliament expressed its belief that 'sustainable growth and employment are Europe's most pressing goals and underpin social and environmental progress' and that 'well-designed social and environmental policies are themselves key elements in strengthening Europe's economic performance'. This placed renewed focus on growth, innovation and employment and encouraged the strengthening of social cohesion and the mobilization of national and community resources in the agenda's economic, social and environmental dimensions. There was a stronger focus on growth and employment, simplification and national ownership via national action plans were perceived as the key elements to relaunch the Lisbon Agenda.

To what extent this represented a rethinking of the ambitions and objectives of economic growth, however, remains questionable as the imperative remains that of economic progress. It may be interpreted as a widening of economic activity to include social and environmental actions. That is an important iteration in its own right, yet does not necessarily change the underlying values and interests in the EU policy framework. This is reflected in the Commission's mid-term review of the Lisbon Agenda which promoted three main objectives: (i) to expedite a stronger focus on 'rigorous prioritization' where it proposed to stress economic growth and employment through a Partnership for Growth and Jobs, this explicit economic agenda was to be supported by an action plan at the EU level and in national action plans by the constituent member states; (ii) active support for reforms in the national member states with a wider political ownership by social partners and citizens was expected so as to take forward the necessary reform processes; and (iii) the simplification, clarification and simpler reporting on progress towards the Lisbon Agenda.

In its first annual progress report on the Lisbon Agenda in January 2006, the Commission defined four priority areas where more action was needed. Next to repeated calls for more investment in education and research, more support for small to medium enterprises (SMEs) and higher employment rates, the Commission took on board one new area, which had hitherto not been a part of the Lisbon Agenda – the need to define a common EU energy policy. A second progress report, adopted in December 2006, concluded that the Commission had delivered on around 75 per cent of the actions it presented as the Community dimension of the Lisbon Agenda, such as the services directive being adopted, progress was being made in financial services and the 7th Research Framework Programme being agreed. 'Unfinished' issues included the portability of pensions, full liberalization of the energy and transport sector and the renewal of the EU's intellectual property system. Four priority areas for action were defined, including: further investment in knowledge and innovation; reducing administrative burdens for SMEs (Better Regulation, www.euractiv.com/en/innovation/better-regulation/article-117503); modernizing labour markets; and energy and climate change. The Commission's December 2007 strategic report, endorsed by leaders at the March 2008 Spring Summit, concluded that the policies defined in the Lisbon Agenda were finally paying off. The report, however, underlined that 'not all member states have undertaken reforms with equal determination' and that reforms in some areas, such as opening up energy and services markets and tackling labour market segmentation, have lagged behind. In March 2008, the European Spring Council endorsed the priorities for the last three years of the Lisbon Agenda, laid out in the Commission's strategic report, and in autumn 2008 member states were expected to present a second round of National Reform Plans, based on the revised integrated guidelines.

What is clear from this unfolding of the Lisbon Agenda is the centrality of economic priorities and the emphasis on growth and development. While the parallel agendas relating to social cohesion, environmental responsibility and territorial equity appeared in the first iteration of the Lisbon Agenda they have become at best marginal and secondary. The implications of this very conventional approach to establishing economic and environmental priorities are very clear – the environment was the conduit to securing economic growth ambitions. In effect, a market-based framework was being promoted, involving and resting on a low ecological consciousness – the Lisbon Agenda, de facto, represented 'business as usual'.

Changing contexts: Ideological and political development

Consideration of the potential of the ecosystem approach to the natural environment and its assertion of the need to accommodate wider, more broadly based and socially constructed values and potentials in the marine context does not take place in a vacuum. It is important to acknowledge at the outset that any discussion about the nature of intervention involves a complex of state, market and civil interests. Here the context established by the market economy needs to be considered. Following Milonakis and Fine's (2009) informed discussion of the evolution of economic theory, it is possible to absorb the significance of the powerful extension of market economic thinking and policy. Based on the pursuit of profit, the reliance on the pricing signals and values, and the assertion of private property rights and market economics form the intellectual, political and practical context to any discussions about the marine environment. In essence, a market economic context prescribes the social construction of the natural environment – it places specific values and invokes assumptions about its use, exploitation and management.

The observations by Giddens (1998), for example, in reviewing the constituent elements of the political manifestos in the UK since 1945, are appropriate and helpful. In assessing the individual contributions to political discourse by the traditions associated with social democracy, neoliberalism and the (so called) Third Way, he identifies an iterative development of political thinking and policy implementation. Here, for example, there have been changing attitudes to the form of government intervention and its relationship with market activities and institutions. Yet, despite arguing that political ideas have changed over this period, Giddens (1998) notes that all of these political ideologies were based on a low ecological consciousness and an emphasis on promoting economic growth. What differed was the means of securing the economic ambitions – either through state-led initiatives and public expenditure, market-based and business led measures or partnerships – but the common theme was that of viewing the natural environment as a source of economic and material well-being for corporate, government and community purposes.

The development of North Sea oil and gas in the early 1970s is a case in point. The early history of North Sea oil and gas exploitation demonstrating the growing importance of the new energy source, which was primarily managed at national levels. In the UK, for example, new legal frameworks were required to establish offshore exploration and production zones. The emphasis was on bringing the oil and natural gas resources onshore to secure the economic, employment and inward investment benefits of this new resource. The offshore–onshore relationship involved the construction of new reception facilities, pipeline landfalls and the designation of areas to facilitate the construction of the offshore platforms and exploration rigs. In parallel, certain localities – Aberdeen and the north east of Scotland, Orkney and Shetland – became centres for offshore exploration and development (Lloyd and Newlands, 1993). Not surprisingly, the development of the offshore marine resource, the new technologies involved in exploiting the new carbon economy, and the large scale of developments required to connect and process the resources, led to a number of potential onshore impacts with widespread environmental implications associated with the construction of offshore installations, service bases, terminals and pipelines for the oil and natural gas (Lloyd and Paget, 1982).

In the early stages of this new marine industry, this required positive action by the government, reflecting its established social democratic values, to protect the wider public interest in the natural and marine environment in the face of the economic growth imperative. As a consequence, a number of land-use planning innovations were put in place to create a strategic context for the developments taking place to secure economic growth. This was intended to address the tensions and contradictions that would likely result from different articulations of national and local public interest (Rowan-Robinson et al, 1989). It was also a response to balance the relationship between the economic advantages to be derived from the offshore development of the marine resource and the associated onshore social and environmental impacts. The specific forms of intervention to mediate these different facets of the natural and marine environment was subsequently critically acclaimed and became an integral part of the UK statutory land-use planning system (Purves and Lloyd, 2008). The primary focus, however, remained on the exploitation and development of the marine resource.

This example is even more powerful in its illustration of the departure from the premise of Giddens (1998) that in the late 1990s there was a rejection of neoliberal economics – towards a synthesis of social democratic thinking based on the notion of market failure, together with the neoliberal viewpoint based on the idea of government failure concerned with recasting economic values. Yet, as Jenkins (2006) argued, the so-called synthesis was in actuality a new statement of neoliberal values and priorities. In effect then, environmental considerations have resided on a market based perspective of the value of the environment to society.

Time does not stand still and, notwithstanding the powerful influence of market economics and business values on public policy, there are countervailing pressures beginning to emerge. In this context the work of Judt (2010) is important. In a powerful polemic, he offers a critique of the pursuit of material self-interest and the concomitant erosion of a sense of collective purpose. Indeed, he argues that the materialistic and selfish quality of contemporary life is not inherent in the human condition. Much of what appears 'natural' today dates from the 1980s – the obsession with wealth creation, the cult of privatization and the private sector, the growing disparities of rich and poor. And above all, the rhetoric that accompanies these: 'Uncritical admiration for unfettered markets, disdain for the public sector, the delusion of endless growth' (Judt, 2010, p2). He calls for a fundamental rethinking of our value systems, our state-market–civil relations, institutional designs and a rebalancing of public and private interests. This is an important clarion call to resurrect the primacy of the collective interest and the appropriate management of common property resources such as the marine environment.

There is further evidence of a growing recognition of the need to rethink societal understandings of the natural terrestrial and marine environments. The Stern Report (Stern, 2006), for example, put forward a considered review of the impacts and risks arising from uncontrolled climate change, and of the costs and opportunities associated with alternative actions to tackle it. It raised important questions for the future role of spatial regulation and planning in securing more appropriate arrangements for the natural environment. Essentially the Stern Report was a direct response to the environmental consequences of economic growth with high material consumption and production of goods and services that resulted in natural resource exploitation and environmental degradation, and to global warming, greenhouse gas emissions and deleterious ecological impacts. Such mainstream advocacy is not isolated. Following the extreme flooding in England in 2007, attention was drawn to the need to provide appropriate regulatory, institutional and planning arrangements to both mitigate and adapt to the uncertainties and impacts of such occurrences (Pitt Review, 2008).

Indeed, this line of reasoning can be extended to promoting the green proofing of government policies and actions, as in seeking to facilitate economic recovery (Bowen et al, 2009; New Economics Foundation, 2009). Moreover, this reasserts the importance of addressing climate change and devising sound environmental policies as relatively more important than economic recovery. Governmental and community investment in green infrastructure, for example, such as flood protection and coastal defence schemes, are advocated as a responsible way forward.

Here too, attention can be drawn to the contribution of the European Commission on the Measurement of Economic Performance and Social Progress to advocating a new social construction of the natural environment. In 2008, President Sarkozy of the French Republic commissioned a study to review the state of statistical information about social and economic matters in the modern world. The study was undertaken by the Commission on the Measurement of Economic Performance and Social Progress and its report offers a radical insight into the importance of statistical indicators that are used to inform government priorities and public policies. It concludes that we need improved metrics to allow us to better design more appropriate policies and measures to address the dynamics of current times. Here, the report makes the important distinction between an assessment of current well-being and an assessment of sustainability.

The Commission argues that society should change its metrics in order to shift from the measurement of economic production to measuring well-being and sustainability. The advocacy of a new system of measurement that incorporates a more sensitive appreciation of social well-being may prove to be highly significant. It will bring questions of equity – both intra- and inter-generational – to decision-making processes. Environmental considerations will command more respect and quality-of-life measures will better allow for perceptions and psychological positions to be taken into account. It will also recast notions of the spirit and purpose of the land-use planning system. At the end of the day, land and marine spatial planning can then shrug off its instrumental niche as a regulator of development and reassume its potentially transformative role in managing social, economic and environmental change in the wider public interest.

The Lisbon Agenda thus represents a set of values and thinking of a specific time. It articulated the prevailing ideas asserting the primacy of economic growth and development. It effectively placed the natural environment and resources as a secondary priority – although the rhetoric suggested otherwise. This is revealed to a certain extent by the questions associated with sustainable development which was achieved in the context of the Lisbon Agenda (Hales, 2000; Connelly, 2007). In effect, sustainable development relied on conventional measures that tended to reflect the interests of private sector stakeholders, interests and values (Bunce, 2009). Now, two processes of change are evident. On the one hand, the economic context to established policies and agendas for action has been transformed. There is even greater pressure on the political ambitions for economic growth and this brings the concomitant threat that the natural environment will come under greater exploitation and development pressures. On the other hand, however, there appears to be a shift in thinking towards seeking an alternative environment-sensitive set of values. Here, EA has the potential to offer itself as a countervailing programme of action to better reflect wider social and community values around the natural environment. What is required, however, is a deliberate opportunity to debate, explore and critically consider issues relating to the marine environment (IPPR, 2008). The advocacy of the need for such a political space is proposed as it would allow for a more balanced and reasoned discourse which would otherwise be obscured by conventional political processes. EA must not be seen simply as an alternative approach but as a political process that seeks to transform the management of both natural terrestrial and marine environments.

Conclusion: EU development and maritime policy

In reviewing relationships between EU constitutional and political development on the one hand, and the evolution of EU maritime policy on the other, a number of key points emerge, related respectively to EU development itself, the marine environment, related economic and social change, and regional management and planning. The advent of EA in environmental management provides an opportunity to draw together environment and development themes and issues, both theoretically and in practical application.

The second stage of EU evolution, from the 1990s onwards, has been characterized by an accelerating pace of enlargement and accompanying constitutional development exemplified by the successive treaties, a process that appears to be slowing in the face of national political realities. It has also been associated with the complementary attempts to promote frameworks for economic development and promotion of sustainability represented respectively by the Lisbon and Gothenburg Strategies. Contemporaneous with this has been the advent of the Water Framework and Strategic Maritime Framework Directives, together with the Maritime Policy and the reworking of the CFP. The CFP is on the verge of entering a new phase of development, while the

Maritime Policy is just beginning. This EU evolution has taken place against a background of both relatively continuous economic expansion culminating in the major recession that commenced in 2008, coupled with increasing differentiation in regional economic development between the core on the one hand and the vast maritime periphery on the other.

In an environmental context the present phase of the CFP has been associated with unremitting pressure on fish stocks and increasingly determined efforts to limit catching power – the overall result in conservation terms has been failure. A radical new approach is beginning to take shape for the post-2012 stage of development, involving introduction of ITRs and bringing the industry stakeholders in particular more fully into decision-making processes. In contrast to the CFP failure, considerable strides have been made in the consolidation and further development of marine environmental legislation at EU level and related implementation at national level.

In an economic and social context, the emergence of the Maritime Policy is of great political significance, although its application lies largely in the future. This event acknowledges the growing realization of the interdependence of sea uses and related resource exploitation and environmental impacts, not least within the remit of the CFP, which is continually adapting to the need for environmental sustainability. Meanwhile, there are growing problems of economic restructuring and social change associated with the continued decline of maritime communities in the periphery especially. Foremost is the need to diversify away from the fishing industries into other fields, including marine renewable energy generation, and continued investment in offshore oil and gas development.

In terms of regional environmental management and related integration, there is a continuing trend towards this, both within the EU (De Santo, 2010; Meiner, 2010), at national level and with European international organizations and initiatives such as the North Sea Ministerial Conferences and the Oslo and Paris Commission. Significantly, these are highlighted by recent initiatives in core countries to provide a much greater level of integration in marine management, including marine spatial planning. The Netherlands has led in this field for a long time, the Dutch Continental Shelf is recognized as one of the country's planning regions. There are also significant developments in Germany, Belgium and especially in the UK and its devolved administrations (Douvere and Ehler, 2008), including the UK Marine and Coastal Access Act 2009 and the Marine (Scotland) Act 2010.

It must be underlined that the constitutional and related political developments have been of decisive importance in the evolution of the CFP, with its high degree of centralization of decision-making and consequent derogations from national sovereignty, and in environmental law-making, where the directives have become decisively influential in the framing of national environmental legislation. However, not so with regard to the strategies, which cannot be said to have been particularly influential in maritime governance processes which have continued throughout to promote economic development at the expense of natural resources and environment, especially in the case of fisheries.

It is in this context that the relatively new idea of EA to environmental management represents a crucial development in historical terms. Its significance lies in the possibility of establishing an alternative to more conventional approaches to the management of the natural environment. These have tended to be driven by capitalist market values based on exploitation and development for material production of goods and services. The driving forces have been profit oriented and short term in nature. In a number of cases, such as fisheries, this has led to over-exploitation of the natural environment, the exhaustion of natural resources, as well as further issues of pollution, waste and climate change. This conventional market exploitative approach to the management of the natural environment imposes wider social, community and territorial costs on society and longer term economic costs. It results in an effective dysfunctional relationship between prices and values in the natural environment and similarly dysfunctional state–market–civil society relationships as conflicts and tensions arise over the misuse of natural assets. In contrast, EA, despite significant conceptual and operational difficulties from both natural and social science perspectives, is nonetheless held to offer a framework for achieving sustainable development and utilization of marine resources which ensures that people and economic systems are not merely the sources of environmental challenges but also part of the solution.

References

Armstrong, H. and Taylor, J. (2000) *Regional Economics and Policy*, 3rd edition, Blackwell, London

Ballinger, R. C. and Stojanovic, T. A. (2010) 'Policy development and the estuary environment: A Severn Estuary case study', *Marine Pollution Bulletin*, vol 61, pp132–145

Ballinger, R. C., Smith, H. D. and Warren, L. M. (1994) 'The management of the coastal zone of Europe', *Ocean & Coastal Management*, vol 22, no 1, pp45–85

Bell, S. and McGillivray, D. (2006) *Environmental Law*, 6th edition, Oxford University Press, Oxford

Bowen A., Fankhauser, S., Stern, N. and Zenghelis, D. (2009) *An Outline of the Case for a 'Green' Stimulus*, policy brief, Grantham Research Institute on Climate Change and the Environment, London, February 2009

Brinkhorst, J. L. (1999) 'European environmental law: An introduction', in N. S. J Koeman (ed) *Environmental Law in Europe*, Kluwer Law International, London

Bunce, S. (2009) 'Developing sustainability: Sustainability policy and gentrification on Toronto's waterfront', *Local Environment*, vol 14, no 7, pp651–667

Connelly, S. (2007) 'Mapping sustainable development as a contested concept', *Local Environment*, vol 12, no 3, pp259–278

De Santo, E. (2010) 'Whose science? Precaution and power-play in European marine environmental decision-making', *Marine Policy*, vol 34, no 3, pp414–420

Douvere, F. and Ehler, C. (eds) (2008) Special Issue on the role of marine spatial planning in implementing ecosystem-based sea use management, *Marine Policy*, vol 32, no 5, pp759–843

EC (European Commission) (2008) *Green Paper: Reform of the Common Fisheries Policy*, European Commission

EC (2010) https://webgate.ec.europa.eu/pfis/cms/farnet/, accessed 15 August 2010

Ecoports (2010) www.ecoports.org.eu, accessed 10 August 2010

Elliott, M., Fernandes, T. F., and de Jonge, V. (1999) 'The impact of European Directives on estuarine and coastal science and management', *Aquatic Ecology*, vol 33, pp311–321

Giddens, A. (1998) *The Third Way: The Renewal of Social Democracy*, Polity Press, Cambridge

Haigh, N. (1999) *Manual of Environmental Policy: The EC and Britain*, Elsevier Science, London

Hales, R. (2000) 'Land use development planning and the notion of sustainable development: Exploring constraint and facilitation within the English planning system', *Journal of Environmental Planning and Management*, vol 43, no 1, pp99–121

Holden, M. J. (1994) *The Common Fisheries Policy: Origin, Evaluation and Future*, Fishing News Books, Farnham

IPPR (Institute for Public Policy Research) (2008) *Engagement and Political Space for Policies on Climate Change*, IPPR, London

Jenkins, S. (2006) *Thatcher and Sons: A Revolution in Three Acts*, Penguin Books, London

Jordan, A. (1998) 'European Community water quality standards: Locked in or watered down?', *CSERGE Working Paper*, WM 98, pp1–32

Judt, T. (2010) *Ill Fares the Land*, Penguin Books, London

Lloyd, M. G. and Newlands, D. (1993) 'The impact of oil on the Scottish economy with particular reference to the Aberdeen economy', in W. Cairns (ed) *North Sea Oil and the Environment*, Elsevier, Barking, pp115–138

Lloyd, M. G, and Paget, G. E. (1982) 'Resource management and land use planning: Natural gas in Scotland', *Journal of Environmental Management*, vol 15, pp15–23

Long, R. and O'Hagan, A. M. (2005) 'Ocean and coastal governance: The European approach to integrated management: Are there lessons for the China Seas region?', in M. H. Nordquist, J. N. Moore and K. Fu (eds) *Recent Developments in the Law of the Sea and China*, Brill, Dordrecht

Meiner, A. (2010) 'Integrated maritime policy for the European Union: Consolidating coastal and marine information to support maritime spatial planning', *Journal of Coastal Conservation, Planning and Management*, vol 14, no 1, pp1–11

Milonakis, D. and Fine, B. (2009) *From Political Economy to Economics Method, the Social and the Historical in the Evolution of Economic Theory*, Routledge, London

New Economics Foundation (2009) *A Green New Deal: Joined-up Policies to Solve the Triple Crunch of the Credit Crisis, Climate Change and High Oil Prices*, New Economics Foundation, London

O'Hagan, A. M. and Ballinger, R. C. (2009) 'Coastal governance in north west Europe: An assessment of approaches to the European stocktake', *Marine Policy*, vol 33, no 6, pp912–922

O'Hagan, A. M., Ballinger, R. C., Ball, I. and Schrivers, J. (2005) *COREPOINT: European Legislation and Policies with Implications for Coastal Management*, COREPOINT

Page, B. and Kaika, M. (2003) 'The EU Water Framework Directive: Part 2. Policy innovation and the shifting choregraphy of governance', *European Environment*, vol 13, pp328–343

Pitt Review (2008) *Learning Lessons from the 2007 Floods: Final Report*, London

Purves, G. and Lloyd, M. G. (2008) 'Identity and territory: The creation of a national planning framework for Scotland,' in S. Davoudi and I. Strange (eds) *Conceptions of Space and Place in Strategic Spatial Planning*, Spon, London, pp86–109

Rowan-Robinson, J., Lloyd, M. G. and McDonald, D. (1989) 'National Planning Guidelines: Their role in strategic policy making and plan implementation', in R. Grover (ed) *Land and Property Developments: New Directions*, Spon, London, pp132–147

Sherman, K., Alexander, L. M. and Gold, B. D. (eds) (1993) *Large Marine Ecosystems: Stress, Mitigation and Sustainability*, AAAS Press, Washington, DC

Smith, H. D. (1991) 'The application of maritime geography: A technical and general management approach' in H. D. Smith and A. Vallega (eds) *The Development of Integrated Sea Use Management*, Routledge, London, pp7–16

Spalding, M. D., Fox, H. E., Allen, G. R., Davidson, N., Ferdana, A., Finlayson, M., Halpern, B. S., Jorge, M.A. Lombana, A., Lourie, S., Martin, K. D., McManus E., Molnar, J., Recchia, C. A. and Robertson, J. (2007) 'Marine ecoregions of the world: A bioregionalization of coast and shelf areas', *BioScience*, no 57, pp573–583

Stern, N. (2006) *The Economics of Climate Change*, HM Treasury, London

Wise, M. J. (1984) *The Common Fisheries Policy of the European Community*, Methuen, London

Chapter 4

Marine Planning and Management to Maintain Ecosystem Goods and Services

*Chris Frid, Geraint Ellis, Kirsty Lindenbaum, Tom Barker and
Andrew J. Plater*

This chapter aims to consider:

- What constitutes ecosystem goods and services;
- The scale and value of marine ecosystem goods and services in the UK;
- The role of the marine ecosystem processes in delivering valuable goods and services;
- The major pressures on these processes;
- Possible planning and management responses to these pressures; and
- The value of the ecosystem goods and services concept in marine planning and management activities.

This chapter explores the delivery of ecosystem goods and services by the marine environment and the role of marine planning and management in protecting this as a key economic resource. It has been written in the context of increasing threats to the marine environment and the calls for 'new forms of governance' (for example, by the European Commission, EC, 2007) in science–policy relations that are required for the sustainable management of the world's marine environments (Plasman, 2008; Fritz, 2010), by seeking to translate the innate value of marine ecosystems in terms of utility to human society. This chapter begins with a consideration of what constitutes ecosystem goods and services. It then considers the scale and value of marine ecosystem goods and services in the UK. The role of the marine ecosystem processes in delivering valuable goods and services is briefly described and the major pressures on them are itemized. The possible planning and management response to these pressures and the use in protecting marine ecosystem services is then discussed.

Introduction

The UK's Marine and Coastal Access Act 2009 has introduced a new institutional regime for managing the country's marine environment, with the explicit objective of contributing to the 'achievement of sustainable development' (Section 1). Although it is debatable whether this fully reflects an ecosystem approach to management (Arkema et al, 2008), it does imply a strong emphasis on protecting the integrity of the marine environment and ensuring this is appropriately balanced with economic priorities and the other needs of human society. A fundamental component of this and other marine management regimes is the ability to understand, value and protect the critical elements of the marine environment that contribute most to the functioning of wider socio-economic systems and the integrity of marine and global ecosystems. One approach to developing an appropriate evidence base has been the concept of ecosystem services.

For this purpose, an ecosystem can be defined as 'a biological community of interacting organisms and their physical environment' (Oxford English Dictionary online – www.askoxford.com), thus the key elements are the biological organisms, the physical (and chemical) environment and the interactions between these components.

The ecosystem can be seen to deliver a variety of goods (such as food resources) and services (such as waste assimilation and treatment) that may have value to human society, above and beyond just maintaining ecosystem functioning. Indeed, ecosystem services can be positively linked to the critical issues of security, health, stable social relations and the basic needs (shelter, food, and so on) required for human survival (MEA, 2005). It must be remembered that these are consequences of the activities of the organisms present, their life processes and their interactions with the physical, chemical and biological systems within which they live. Ecosystems are abstract, human imposed units and do not exist to, nor strive to, deliver any particular suite of goods and services. Therefore, the range of ecological goods and services provided by any ecosystem, be it a rock pool or the North Atlantic, is entirely dependent on the species present in the system and their life processes (Table 4.1 provides a list of goods and services supplied by the UK marine environment).

In some cases it is possible to measure the ecosystem goods and services directly but in most cases it is not, and reliance must be placed on predictive tools, such as modelling, which are based on measurements of organism abundance and process rates. For both ecological and practical reasons, consideration of the delivery of ecological functions must, therefore, proceed from an understanding of the identities and roles of the organisms comprising the biological assemblage (Bremner et al, 2006).

It should be noted that many of the key ecosystem processes, including much of the primary production of food, the breakdown of pollutants, and the cycling of nutrients, are carried out by microbes. There is very limited knowledge of the ecology of the microbial system in the marine environment (or on land). Genetic markers for key enzymes tell us something about the availability of, for example, nutrient processing capacity but not in what organisms the enzymes reside or how the organisms might respond to changes in other parts of the ecosystem. However, while microbes are the actual 'doers' they are strongly influenced by the actions of the larger organisms in the system where more understanding does exist (Aller, 1988; Blackford, 1997; Howe et al, 2004).

A critical attribute in relation to the interaction with human societies is that many natural ecosystems have an inherent ability to show recovery following disturbance. This is only possible for impacts that are below a certain threshold. Disturbances that exceed that threshold force the system into a new configuration. The ability to recover is a valuable property. It means systems can tolerate the impacts of human activities, up to a certain point, without an alteration to the system's dynamics and the delivery of ecosystem processes, so that they may be able to assimilate waste products from human activity or endure the human harvesting of resources. The ability of a particular ecosystem component to resist an impacting activity (resistance) and its ability to subsequently recover (resilience) varies and is dependent on the biology of the species present. This inherent ability to respond to externally and internally forced change and then recover often lies in perhaps unseen species or roles that then come into play as a result of change or disturbance. This exemplifies the unseen, or even unknown, contingency (or redundancy) that lies within an ecosystem that perhaps only operates during times of stress, yet this contingency is a critical element that helps prevent collapse and provides resilience to stresses.

While we acknowledge the 'ecosystem' as a human construct, as well as the benefits of regarding the ecosystem as an open but coherent system, it is important that marine ecosystems are not considered as operating to deliver any particular suite of goods and services. However, in adopting the concept of ecosystem goods and services, it is enticing for management to focus on the delivery of these for the human health and well-being. It is important, therefore, to emphasize that the value of the ecosystem approach is that it serves as a mechanism for conceptualizing nature, its complexity and inherent variability. More significantly, it captures the critical issues of connectedness, flows of energy and matter, process and response, sensitivity and resilience to change and disturbance, and the capacity for self-regulation. These issues are highly significant in understanding marine ecosystems, laying the foundations of any management activity.

For the purposes of this chapter, the marine ecosystem can be divided into six components, ignoring for the moment the human presence in such ecosystems (such as their function as predators or polluters):

1. Plankton (primarily microscopic organisms, but also including larger organisms that float in the water, such as jellyfish and comb jellies);
2. Benthos (organisms that live on or in the sea floor);
3. Fish;
4. Marine reptiles and mammals (such as sea turtles, dolphins, seals);
5. Seabirds; and
6. Sea floor habitats.

This typology has previously been used both to ease consultation with stakeholders and to make the modelling of ecosystem responses to management more tractable (Paramor et al, 2004). It has also formed the basis of the recent International Council for the Exploration of the Sea (ICES)/OSPAR work to develop integrated metrics of system health (ICES, 2005, 2006). In these examples, criteria were developed to assess the importance of individual habitats and species. The principle criteria were based on economic or societal importance, including those species that were exploited commercially (either directly by harvesting or indirectly through ecotourism). This is a critical principle of the ecosystem services approach, in that it attempts to capture, in tangible forms, the value of ecosystem function, not necessarily for its own intrinsic value but in terms of its contribution to human survival or prosperity (in other words, anthropogenic instrumental value, rather than any intrinsic value or non-anthropocentric value, see Turner et al, 2002). The purpose of this is not to place an oversimplified monetary value on different elements of marine ecosystems, but to act as a decision tool, acknowledging the limitations in doing such approaches (for example, Sagoff, 1998; Clark et al, 2000).

What are ecological functions?

Conceptualizing and understanding the importance of ecosystem goods and services is non-trivial but it is of critical significance that the importance of healthy and diverse marine ecosystems is made tangible to the non-specialist. Ecosystem goods and services might, therefore, be regarded as a transdisciplinary communication tool through which marine scientists are able to engage in effective discourse with social scientists, management practitioners, regulatory authorities and a spectrum of stakeholders. The value of this tool overcomes disciplinary boundaries and the need for all concerned to understand the natural science foundations of marine ecosystems where a certain level of ambiguity requires specialist knowledge. The concept of ecosystem 'function' demonstrates this problem for the non-specialist:

● Function – the role an organism plays in the ecosystem.
● Function – processes by which the ecosystem operates (flows of matter and energy).
● Function – capturing how the ecosystem provides society with goods and services.

Marine scientists can readily grasp the context in which this term is used, and hence the meaning, while the same cannot be said for the practitioner implementing policy. The same applies to terms such as redundancy and contingency which are key elements in capturing the resilience of marine ecosystems.

Figure 4.1 The distribution of marine landscape features in UK seas

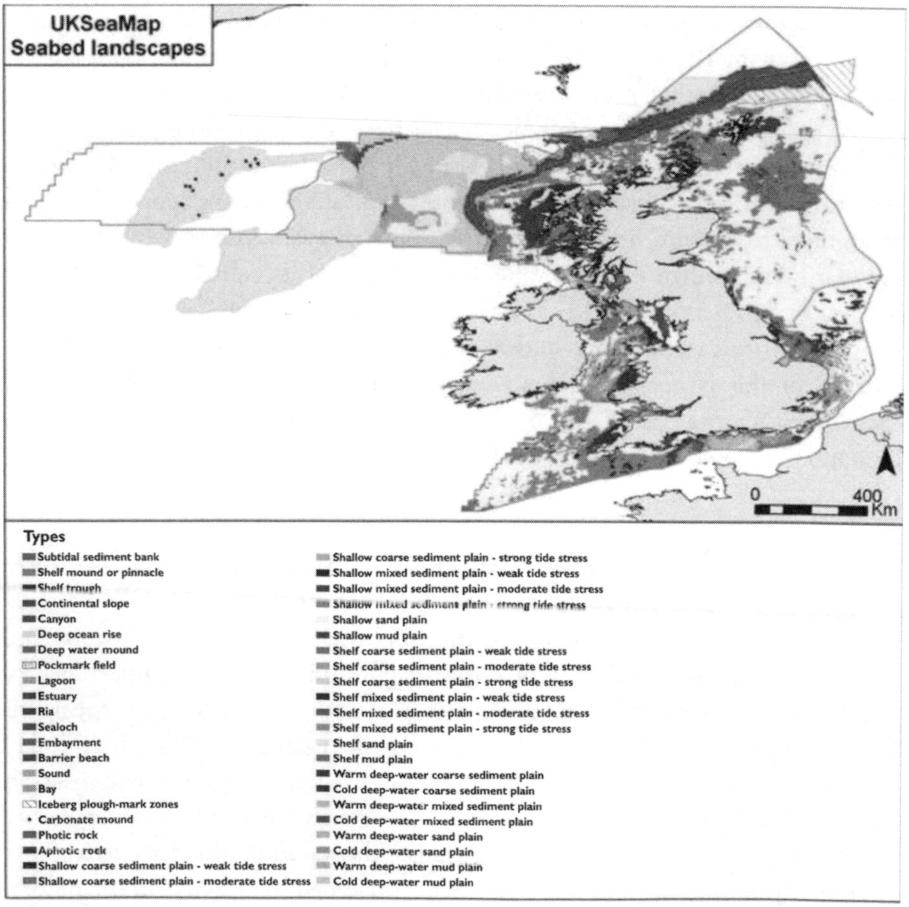

Types

Subtidal sediment bank	Shallow coarse sediment plain - strong tide stress
Shelf mound or pinnacle	Shallow mixed sediment plain - weak tide stress
Shelf trough	Shallow mixed sediment plain - moderate tide stress
Continental slope	Shallow mixed sediment plain - strong tide stress
Canyon	Shallow sand plain
Deep ocean rise	Shallow mud plain
Deep water mound	Shelf coarse sediment plain - weak tide stress
Pockmark field	Shelf coarse sediment plain - moderate tide stress
Lagoon	Shelf coarse sediment plain - strong tide stress
Estuary	Shelf mixed sediment plain - weak tide stress
Ria	Shelf mixed sediment plain - moderate tide stress
Sealoch	Shelf mixed sediment plain - strong tide stress
Embayment	Shelf sand plain
Barrier beach	Shelf mud plain
Sound	Warm deep-water coarse sediment plain
Bay	Cold deep-water coarse sediment plain
Iceberg plough-mark zones	Warm deep-water mixed sediment plain
Carbonate mound	Cold deep-water mixed sediment plain
Photic rock	Warm deep-water sand plain
Aphotic rock	Cold deep-water sand plain
Shallow coarse sediment plain - weak tide stress	Warm deep-water mud plain
Shallow coarse sediment plain - moderate tide stress	Cold deep-water mud plain

Source: Connor et al (2006). Copyright JNCC and UKSeaMap funding partners (2006)

Ecosystem functioning is, in essence, the processes and activities that keep a system working (Bolger, 2001) and has numerous definitions including: 'The activities, processes or properties of ecosystems that are influenced by its biota' (Naeem et al, 2004). Other definitions include nutrient recycling (Biles et al, 2002; Naeem and Wright, 2003), the flow of energy and materials through the biotic (living)/abiotic (non-living) components of ecosystems (Diaz and Cabido, 2001), a combination of ecosystem processes and ecosystem stability (Bengtsson, 1998) and the sum total of these processes (Virginia and Wall, 2001).

The environment has different types of value to humans (such as economic use and non-use) which have been categorized by various authors to make economic value estimates (see Table 4.1 for categories). Not all of these values are relevant to the marine environment and the number of categories considered in this chapter has been further reduced from those used elsewhere to bring together functions that are delivered by the same ecological process to reduce duplication and to reflect gaps in knowledge. Table 4.1 shows how the limited number of categories that were used and how they were mapped to other valuation categories schemes.

The scale and value of marine ecosystem goods and services in the UK

The total marine Exclusive Economic Zone (EEZ) claimed by the EU extends to $11,447,075 km^2$. The three largest national contributors are Ireland ($890,688 km^2$, 7.8 per cent), the UK ($867,000 km^2$; 7.6 per cent) and Spain ($683,236 km^2$, 6.0 per cent) (http://earthtrends.wri.org/country_profiles/index.php?theme=1& rcode=2). Portugal is the only other EU nation with both oceanic and coastal seas. Therefore, the UK has both a significant proportion of the total European seas and is responsible for a large part of the oceanic component of the European EEZ and would, therefore, be expected to derive a substantial level of services and wealth from this territory. While it is impossible at this time to partition the goods and services provided by the European seas to any particular area there is a good basis for considering the UK's marine waters to be important in the European context.

The global ecosystem has been estimated to provide around $US\$33 \times 10^{12}$ to the global economy of which some $US\$21,659 \times 10^9$ is contributed by aquatic systems (Pilskaln et al, 1998). The majority of this is derived from coastal systems which deliver 16 times more goods and services than oceanic areas hectare for hectare. This is a reflection of both their greater proximity to population centres and their higher rates of biological processes such as productivity and nutrient cycling. Indeed, when we refer to the management of marine ecosystems in the

UK and Europe, we are concerned less with the unexplored deep sea and much more with shallow seas on the continental shelf generally less than 200m in depth. These seas are seasonally stratified and have about four times the biological productivity of the abyssal ocean owing to being in the photic zone, and are heavily impacted by humans in the form of eutrophication and pollution.

Table 4.1 Categories of ecosystem goods and services as defined by Costanza et al for their global environment, by Beaumont and Tinch for the UK sea floor and Frid and Paramor for UK waters

Costanza et al (1997) ecosystem service categories[1]	Beaumont and Tinch (2003) ecosystem service categories	Frid and Paramor (2006) ecosystem service categories
1. Gas regulation	6. Gas and climate regulation	Gas and climate regulation
2. Climate regulation		
3. Disturbance regulation	9. Disturbance prevention and alleviation[A]	
8. Nutrient cycling	5. Nutrient cycling	Nutrient cycling
9. Waste treatment	7. Bioremediation of waste	Waste treatment
11. Biological control		
12. Refugia[B]	8. Biologically mediated habitat	Habitat functions
13. Food	1. Food provision	Food and material provision[C]
14. Raw materials	14. Raw materials	
15. Genetic resources		
16. Recreation	3. Leisure and recreation	
17. Cultural	10. Culture and heritage	
	11. Cognitive values	
	12. Option use value	Biodiversity in support of societal values[D]
	13. Non Use values – Bequest and Existence	
	4. Resilience and resistance	

A Relates primarily to the flood prevention role of intertidal habitats which are not considered in this report.

B The supporting text in Costanza et al (1997) makes it clear that this function is provided by the habitat.

C Provision of raw materials other than foods are small in the UK and the issues are not different from those surrounding food provision.

D While this issues are distinct economic categories, the relationship to biological diversity are the same so that they are considered together.

For the purposes of this chapter, the provision of ecological goods and services are considered under six categories:

1. Atmospheric gas assimilation and climate regulation;
2. Nutrient recycling;
3. Waste assimilation capacity;
4. Habitat functions;
5. Food provision; and
6. Biodiversity for society (such as recreation: angling, bird watching, boating, coastal visits and diving, existence and cultural values).

These six categories cover the key ecological functions delivered by the UK marine ecosystem (Table 4.1).

Atmospheric gas assimilation and climate regulation

The oceans play a major role in global climate regulation both because of their massive heat capacity, which buffers variations in temperature, and because ocean currents move heat around the globe (Bigg et al, 2003). The temperate nature of the UK climate compared to that of similar latitudes in the north-east US, for example, attest to the warming effects of the 'Gulf Stream'. The economic value of climate stability has recently been underlined by the UK Government Treasury report that estimated that the cost of stabilizing carbon dioxide (CO_2) emissions could be as much as 1 per cent of global Gross Domestic Product (GDP), while the costs of a temperature increase of 5–6°C would be 5–10 per cent of global GDP (Stern, 2006).

The oceans also play a role in atmospheric processes through the chemical exchanges that occur across the air–sea interface (Reid et al, 2009). The atmosphere is the major source of some contaminants that enter the oceans, but the oceans also recharge the atmosphere with oxygen and absorb CO_2. While some of this is a purely physical–chemical process, most is mediated by biological systems (Liss, 2002; Fowler et al, 2009). For example, the flux of oxygen to the atmosphere is derived from marine plant photosynthesis. However, some of the chemicals absorbed into the ocean are sequestered into deep-sea sediments and so are effectively removed from the living, dynamic, part of the global system. Planktonic organisms, such as coccolithophorids and radiolarians, use the dissolved CO_2 to produce shells of calcium carbonate. When the organisms die, their shells, being large and heavy, rapidly sink to the sea floor taking the carbonate (CO_2) with them. Some marine plants also liberate gases that are important in stimulating cloud formation. The cloud cover of the planet is important because it reflects some incoming radiation, traps some heat (and so buffers the variation at the planet surface) and alters patterns of rainfall. Without the living marine ecosystem, the Earth's atmospheric systems would be less hospitable to life on land and subject to much greater threats than are currently being debated as a result of climate change.

Nutrient recycling

Nutrient availability strongly affects productivity in the marine environment (Hecky and Kilham, 1988). Nutrient concentrations in the water column follow a strong seasonal cycle with blooms of phytoplankton (microscopic plants in the water column) in the spring and autumn. These occur as the surface and deep waters are mixed during the winter storms, bringing nutrients from the deeper areas up to the surface.

The ecosystem of the continental shelf typically receives half the nutrients it requires for primary production from the sediment (Pilskaln et al, 1998). These nutrients are derived from the dead material that accumulates on the sea floor (Gooday, 2002). This organic material is decomposed by bacteria which releases the nutrients back into the water column through chemical diffusion processes which may be enhanced by sediment irrigation activities performed by benthic communities (Aller, 1982). The chemical relationship between the water column and sea floor is tightly coupled (Nixon, 1981; Blackford, 1997; Gowen et al, 2000).

Capacity for waste assimilation

The marine environment represents an important sink for waste material discharged directly into the sea via outfalls (point source discharges) and via rivers/estuaries and the atmosphere (diffuse source discharges). These wastes may be assimilated by living organisms (Sly, 1989; Van Dover et al, 1992; Prince, 1993) or may become locked in the sediment (Horowitz and Elrick, 1987; Perez et al, 1991; Valette-Silver, 1993; Rees et al, 1998).

Non-biodegradable substances, such as heavy metals, and some synthetic substances, such as pesticides and polychlorinated biphenyls (PCBs), may be dispersed by water movement and dilution (Somerville et al, 1987; Swannell et al, 1996), or may become concentrated in living organisms. This can lead to a magnification of these substances up the food chain, causing health concerns for top predators including man (Phillips and Rainbow, 1989; Leah et al, 1991; Galay Burgos and Rainbow, 2001; Rainbow et al, 2004)

Biodegradable wastes, such as food wastes and the organic components of sewage, in addition to being dispersed (Hughes and Thompson, 2004), are also subject to breakdown and ultimately decomposed to carbon dioxide and inorganic nutrients.

Habitat functioning

Habitats provide the 'living space' required by organisms. They provide sites for feeding, breeding, sheltering from predators and natural disturbances and, thus, are a prerequisite for the provision of many other goods and services.

Most, but not all, definitions of habitat include both the abiotic and biotic factors that characterize the location. These factors may include physical structures such as sediments, reefs and biological structures. The latter may include worm tubes, coral reefs and the tentacles of jellyfish.

The availability of the correct 'habitat' is therefore a requirement for an organism to survive, grow and reproduce. This critical role is emphasized in the US by the provisions of the Magnuson-Stevens Fishery Conservation and Management Act (1996) that requires regional fisheries councils ensure effective protection of 'essential fish habitat' (for example, the nursery and feeding habitats of commercial fish species) as part of their commitment to ensuring sustainable fisheries.

The EC Habitats Directive (1992) requires member states to provide protection for a limited number of marine habitat types. The habitat types were selected primarily for their societal value rather than to ensure delivery of key ecosystem functions.

Food provision

The marine environment is a readily available source of food for human consumption. In 2008, the UK fleet landed 588,000 tonnes of sea fish with a total value of £629 million and supported 12,761 fishermen (Marine Fisheries Agency, 2008). This is in addition to the regional economic effects of the fishing industry and the critical support it provides to some coastal communities.

Fisheries exploit a diverse range of species including:

- bottom-feeding shellfish such as whelks and *Nephrops* (scampi or langoustine);
- shellfish that filter their microscopic food from the water, such as mussels, cockles and scallops;
- fish such as plaice, sole, cod, haddock and skate, that feed on bottom-dwelling worms, crabs, sandhoppers and clams; and
- fish that feed on other fish and other swimming and floating organisms in the water body (herring, mackerel).

Table 4.2 The quantity and value of UK fish and shellfish landings 2000–2004

	Quantity (thousand tonnes)					Value (£ million)				
	2000	*2001*	*2002*	*2003*	*2004*	*2000*	*2001*	*2002*	*2003*	*2004*
Demersal	301.0	270.3	242.5	202.7	231.1	302.3	281.1	257.2	219.9	223.5
Pelagic	311.8	323.7	305.3	292.9	290.9	78.5	114.2	114.4	114.5	105.8
Shellfish	135.4	143.8	137.6	144.0	131.7	169.5	179.1	174.0	193.9	183.7
Total Fish	748.1	737.8	685.5	639.7	653.7	550.3	574.4	545.6	528.3	513.0

Source: Marine Fisheries Agency (2005)

These fisheries all provide food directly for human consumption. Smaller fish such as sandeels, Norway pout and blue whiting are harvested and converted into fish meal which is used to make animal feeds, including feeds for use in fish farms. Therefore farmed salmon, trout and sea bass derive part of their nutrition from the food provision service provided by the UK seas.

In addition to the harvesting of sea fisheries resources, further revenue and employment is also created through the fish processing industry, retail sales, and exports, with fish processing employing approximately 18,180 people and 1300 fishmongers. Many of these jobs are concentrated in remote communities.

Landings by the UK fleet into the UK have remained around 460,000 tonnes over the last five years. Demersal species (fish that feed on organisms in/on the sea floor) represents 34 per cent of total landings in terms of quantity and 41 per cent in terms of value. Pelagic species (fish that feed on organisms in the water column) account for 39 per cent of landings by quantity but only 16 per cent by value. Shellfish account for 27 per cent of landings by quantity and 43 per cent by value (Table 4.2).

Societal value of biodiversity – recreation, angling, boating, bird watching, coastal visits and diving, existence and cultural values

Recreation has been one of the fastest growing economic sectors in the UK in the last 50 years (Pugh and Skinner, 2002). Much of the marine and coastal recreation is centred on wildlife and scenery, but the level of dependency varies. Ecotourism activities are usually highly dependent on one or a few species, typically cetaceans (whales and dolphins), birds and/or seals. Recreational sea anglers tend to target a limited number of fish species. Many other sectors concentrate, but not exclusively, their activities in areas that are perceived to be 'scenic' or natural, such as scuba-diving, coastal visits and recreational cruising/

sailing. Generally, coastal visits, such as beach use, do not depend on the presence of specific natural biological systems, but users would be perturbed by some signs of misfunctioning systems, for example, harmful algal blooms or accumulating wastes.

A recent estimate of the total net value of marine leisure and recreation in the UK was £11.77 billion (Pugh and Skinner, 2002). This value included income from bird, cetacean and seal watching, aesthetic value, and the indirect value of tourism associated with commercial fishing activity. The importance of these activities, measured as the numbers involved and their commitment to the activity, is shown in Table 4.3.

The UK seas form a major part, around 7.6 per cent, of the European Seas and the economic benefits of the marine environment to Europe are, to a large extent, dependent on the health and status of the seas of the Ireland, UK and Spain. Thus, delivery of clean, healthy and sustainable European marine waters is dependent, in large part, on activities in the UK sector (Frid et al, 2003).

The UK's marine area covers 867,000km² (335,000 square miles) which is more than three times the UK land area, or to put it another way, more than three quarters of the UK's total area is sea. Yet, until recently, this area has not been subject to a system of comprehensive and holistic management regime.

The scale of marine-dependent activities is significant and includes:

- The UK fish and shellfish catching industry which lands over £540 million in catches each year, resulting in between £800–1200 million of economic activity in the UK (Marine Fisheries Agency, 2005);
- UK recreational anglers spend around £1 billion per year on their sport (Drew Associates Ltd, 2003);
- Offshore oil, gas and aggregate extraction is worth more than £20 billion per year (www.og.dti.gov.uk/information/bb_updates/appendices/Appendix7.htm); and
- Offshore wind-power installations. With a UK target of 10 per cent of energy generation from renewable sources by 2010, considerable growth in the offshore renewables sector is expected and this will be stimulated by government investment, via the Climate Change Levy, of more than £1 billion per year by 2010 (www.dti.gov.uk/energy/sources/renewables/policy/offshore/page22500.html).

In addition to these, the marine environment potentially provides an important resource for bio-prospecting although this has yet to be fully explored.

In addition to these direct uses of the marine environment, the seas provide many indirect benefits:

- They are a major reservoir of biological diversity with more than 44,000 species recorded (Defra, 2002);
- They are a major store of the greenhouse gas carbon dioxide (CO_2) and assist in regulating the Earth's climate; and
- Organisms in the sea play a vital role in nutrient recycling, returning nitrogen, phosphorus and sulphur to the biologically active part of the global ecosystem.

Table 4.3 Recreational participation and club membership levels

Activity	Membership of governing body	Estimated popularity	% users who have membership
Sub Aqua	52,247 members (British Sub-Aqua Club) 51,700 members (Professional Association of Diving Instructors)	Approximately 120,000	83%
Angling	221,699 (National Federation of Anglers) 35,000 (National Federation of Sea Anglers)	2.5 million	0.1%
Ornithology/ Bird watching	1 million (Royal Society for the Protection of Birds)	2 million	50%

Source: UK CEED (2000)

Table 4.4 Turnover and 'value added' by the marine sector in the UK economy (revalued to 1999 prices, all £ million)

Sector	1994–1995 (£ million)		1999–2000 (£ million)	
	Turnover	Value Added	Turnover	Value Added
Oil and gas	15,295	12,310	20,597	14,810
Leisure	10,129	6859	19,290	11,770
Defence	6762	2703	6660	2531
Business services	6417	1099	4535	1080
Shipping	5007	2317	5200	2400
Shipbuilding	4002	1875	3172	1574
Equipment	3565	1438	2326	1358
Fisheries	2392	822	2447	825
Environment	1380	460	1050	435
Ports	1311	918	1690	1183
Construction	826	231	500	190
Research	645	309	609	292
Telecommunications	460	230	500	190
Safety	336	138	316	129
Crossings	178	100	155	87
Aggregates	168	87	131	69
Education	54	28	49	25
Total	**58,927**	**31,923**	**69,227**	**38,948**

Source: Pugh and Skinner (2002)

Between 1999 and 2000, the contribution of marine-related activities to the UK economy was estimated to be £39 billion, or 4.9 per cent of Gross Domestic Product (GDP) (Table 4.4). In 1994–1995, the estimated contribution was £27.8 billion, or 4.8 per cent of GDP. Excluding tourism, the 1999–2000 figure is 3.4 per cent of GDP. This further confirms the importance of marine activities to the UK economy (Pugh and Skinner, 2002).

Marine ecosystem processes and delivery of goods and services

Having described the overall extent and importance of the services provided by the UK marine environment, with some approximate values, we now move on to consider how these are derived from ecosystems and how these may change over time.

Important ecosystem goods and services

Atmospheric gas assimilation and climate regulation

There are no direct measures of gas fluxes into and out of the UK seas, and the large-scale data provided by satellite and other remote sensing operations provide only information on surface abundance of plant pigments. Given that the rate of gas use/production and other biogeochemical processes often vary between species of plant, it is impossible to make detailed inferences about any changes in status of these ecosystem services.

Although there are no data that provide direct estimates of the flux of gases in and out of the UK seas, it may be possible to calculate some of the important elements of the flux using oceanic models. For example, the climatically active gas dimethyl sulphide (DMS) is produced from gases released by planktonic algae such as *Emiliana huxleyi*. Blooms of *Emiliana huxleyi* can be documented from satellite imagery. Thus it is possible to calculate the production of DMS. Estimates suggest that *Emiliana huxleyi* blooms might be important regionally, but play only a minor role in the global regulation of climate (Balch et al, 1992). There is some indication that the centre of *Emiliana huxleyi* activity is moving northwards.

Nutrient recycling

Nutrient concentrations in the water column follow a strong season cycle. The annual spring phytoplankton bloom is linked to the mixing of the water column during the winter storms. This mixing brings nutrients from the deeper waters up to the illuminated surface waters which allows photosynthesis to occur. Autumn blooms occur when the stable water column produced in the summer breaks down and nutrients are brought up to the surface. This bloom is curtailed by the diminishing levels of light as autumn advances. There is some evidence that the timings of the bloom is being influenced by climate change (Edwards et al, 2004). This will impact food web interactions but it is unclear how these changes will affect nutrient recycling. Evidence from the continuous plankton recorder (CPR) shows that within the phytoplankton, diatoms (microscopic plants that secrete a silica 'shell') are now less abundant, whilst dinoflagellates (small rapidly growing plant cells) are becoming more common. As diatoms take up silica and dinoflagellates do not, this certainly means an alteration in the cycling of silica. Dinoflagellates and diatoms also use different amounts of other plant nutrients, particularly nitrate and phosphate. Thus, we can conclude that the observed changes in the size, timing and composition of phytoplankton blooms will have resulted in altered patterns of nutrient uptake and so altered the cycles. A lack of detailed historic data, however, means that we cannot measure these changes directly.

Although there is information on the biota that contribute to nutrient recycling (for example, burrowing worms), and some measure of their processing rates, there are very few studies that have measured the direct effects of human activities on nutrient recycling (Percival et al, 2005; Trimmer et al, 2005). The effect of long-term changes in the benthos on nutrient recycling processes is unknown.

Fishing has been shown to have no impact on oxygen uptake, denitrification or nutrient exchange in the southern North Sea (Trimmer et al, 2005). In the long term, chemical processes in the upper layers of sediment appeared unaffected by trawling. This may be because any changes in nutrient recycling that are likely to have occurred as a result of fishing had already happened by the time studies began.

Waste assimilation capacity

The seas have a finite capacity to absorb human-derived wastes without showing detrimental changes. The limit is set by the rate of removal (dilution) and the breakdown/assimilation capacity. For organic wastes and nutrients, the key limit is usually determined by biological processes. For other wastes, it is the dilution capacity (Clark et al, 1997). There are many well-documented cases of the adverse impacts arising from excessive waste inputs. However, these tend to be limited to near the region of input. It follows logically that reductions in the size of inputs and the number of sites for waste disposal will reduce these adverse effects and mean that the system will have a great capacity to function naturally.

Hazardous wastes

Large reductions in the input of most hazardous wastes to the marine environment have occurred over recent decades (Defra, 2005). Total inputs via rivers and directly of mercury (Hg), cadmium (Cd), copper (Cu), lead (Pb), Zinc (Zn) and the organic contaminant known as γ-HCH to coastal waters have been reduced by 20–70 per cent since 1990, and atmospheric emissions of the same chemicals have been reduced by 50–95 per cent since 1990. However, concentrations of Hg, Cd, Cu, Pb and Zn in both water and sediments are elevated. Highest concentrations of hazardous substances in biota were found in industrialized estuaries and adjacent areas with a known history of contaminant inputs. High concentrations of Cd were also found in fish livers from the offshore Dogger Bank site.

Discharges of radionuclides from Sellafield have decreased significantly since the 1970s, as a result of various measures linked to what are regarded as acceptable environmental limits. Current aqueous discharges are more than an order of magnitude less than peak discharges in the 1970s (Jackson et al, 2000).

The National Marine Monitoring Programme (1998) report states that fisheries and wildlife do not appear to be in serious decline due to contaminant effects and the concentrations of many contaminants are apparently decreasing (NMP, 1998). However, trace or microcontaminants are still causing demonstrable polluting effects in areas where certain marine discharges are poorly diluted, or where this is a legacy of persistent pollutants in fine-grained sedimentary sinks (Matthiessen and Law, 2002).

Overall, the great advances made in recent decades to reduce polluting inputs will mean that the system's waste assimilation capacity is healthier now than in the recent past.

Sewage disposal

Sewage inputs to the sea have decreased dramatically since the 1980s following the implementation of the Bathing Waters (EC, 2002), Shellfish (EC, 1979) and Urban Waste Water (EC, 1991) Directives. These have resulted in marked improvements in UK coastal waters (Matthiessen and Law, 2002). Sewage inputs (including effluent from treatment works) and fertilizer run-off potentially contribute to eutrophication (the stimulation of plant growth above natural levels, often leading to undesirable effects).

Nutrient input

Direct inputs of nitrogen and phosphorus in the UK have been reduced by 35 per cent and 50 per cent, respectively, since 1990 (Defra, 2005). However, there is no evidence of reductions in the quantities entering coastal waters from rivers, which mainly arise as diffuse inputs of nutrients from land run-off. The input from diffuse sources varies with rainfall and river flow rates and varies in the different regions. Although the coastal waters of southern, eastern and north-western England and the inner Bristol Channel are enriched with nitrogen and phosphorus, the correlation with nutrient input is weak (Defra, 2005). The reduced nutrient inputs are not reflected in a detectable reduction in the winter nutrient concentration in the sea. This reflects the importance of the ocean as a source of nutrients to UK coastal and offshore waters.

Hydrocarbons

Inputs of oil from land-based sources, including refineries and offshore operations, are tightly regulated and the total input from offshore sources is fairly constant (Defra, 2005). Spillage of oil from accidents offshore or from tankers or other maritime accidents is, by definition, irregular and impacts are localized and transient.

Dredged material

Ports, harbours and shipping channels can silt up with sediment. Approximately 25–40 million wet tonnes of dredged material is removed annually from access channels, ports, marinas and harbours and deposited in around 150 licensed disposal sites. Since 1992, there has been a slight increase in the overall quantity of dredged material deposited at sea each year. More than 60 per cent of all dredged material is deposited in the southern North Sea and the Irish Sea. Monitoring has demonstrated that impacts tend to be confined within the boundaries of the disposal sites, and impacts are site specific (Bolam et al, 2006).

Overview

The levels of material disposed of in coastal seas and estuaries in the recent past will have impacted on their ability to deliver key ecological goods and services. Recent years have seen major reductions in the levels of discharges and greater investment in onshore treatment. This will have reduced the degree to which the system is compromised and so has increased the ability to absorb occasional insults such as accidental spillages. However, there remain a number of localized areas, particularly some estuaries, where historical inputs still cause measurable declines in the health of the marine ecosystem.

Habitat functions

No definitive map of the actual distribution of UK sea floor habitats currently exists, making it impossible to assess whether the area of certain habitat types has changed and what they have been replaced with. In some cases we do have evidence of habitat loss (for example, cold water coral reefs, maerl beds, mussel (blue and horse) and oyster beds) and can infer the loss of the habitat provision and other functions associated with them.

One patch of a habitat, for example, muddy sand, is also not identical to another patch of muddy sand in terms of the biological assemblage it supports and the ecological functions it delivers. This variation within a habitat is sometimes referred to as 'habitat quality'. There is evidence that in some areas habitat quality has been impacted, for example at dredge spoil disposal sites, and yet the impacted seabed may still be characterized as the same habitat as the surrounding areas, despite its lower quality. Any change in species composition implies that the area might be delivering different ecological functions from the natural situation (Tarpgaard et al, 2005).

Food provision

Over the past decade, the status of stocks of some key demersal (bottom-living) fish species has deteriorated (ICES, 2004). In contrast, the state of pelagic species (those that live in the water column), such as herring, has improved. In the North Sea, four of the eight main demersal stocks are harvested unsustainably or are at risk of being harvested unsustainably. The cod stock remains at historically low levels and is subject to emergency management measures and a recovery plan from 2005. In contrast, herring stocks have increased and the quotas for *Nephrops* (scampi) in the North Sea have been increased in recent years suggesting that they are exploited sustainably (T. L. Catchpole (Centre for Environment, Fisheries & Aquaculture Science), personal communication).

During the past decade, Irish Sea cod and whiting stocks have declined, raising concerns regarding possible stock collapse. Most demersal stocks in the south-west approaches are harvested outside precautionary limits. The northern hake stock is the subject of a management recovery plan introduced in 2004 that includes a lower Total Allowable Catch (TAC) and technical measures (mesh size restrictions). Haddock and *Nephrops* in the west of Scotland are harvested sustainably but the status of many of the other demersal species are either uncertain or considered to be at low historical levels.

Many factors can cause changes in the abundance and distribution of fishes, including natural variation, biological interactions and human activities. Activities that are known to affect the structure and diversity of fish communities include fishing, changes to habitat quality caused by, for example, pollution, eutrophication and habitat destruction, and the introduction of non-native species (Agardy, 2003). Determining the relative impacts of these various factors is difficult, however, and while some studies have shown a correlation between environmental variables and biological components, there are few cases that prove causal relationships.

Commercial exploitation of fish also has impacts on the wider marine environment (Frid et al, 2006). These impacts include those on the abundance, size and genetic make-up of target species, on seabed habitats and non-target animals such as marine mammals, fish and benthic fauna that are also caught during fishing operations, on the genetic diversity of both species and populations, and on the food web itself. Fishing affects non-target species caught as by-catch, and has caused reductions in large bodied and vulnerable species such as skates and rays. Monitoring programmes to determine the quantity and composition of discarded catches are in place in many UK fisheries. Many of the larger target and by-catch species in the North Sea and Irish Sea are now reduced to less than 10 per cent of their expected abundance without fishing, and the mean weight of fish has declined.

Therefore, the trend in the provision of food from the sea is one of overall decline. This, in turn, has driven the fishing industry to seek new species, exploit new habitats (such as deep water) and fish harder to maintain catches. These changes will have negatively impacted on the supporting ecosystem compromising its ability to deliver goods and services and to support the fisheries.

Biodiversity for recreation, existence and cultural values

No direct assessments of the levels of biodiversity used for recreation exist. Data on biodiversity are collected from national monitoring sites for impact management purposes and often show declines due to human activity. Recreational activities are not universal or uniform in the marine environment and tend to avoid impacted locations as they are less desirable to the public. The available data suggest that there have been significant increases in recreational sea angling, boating, sport diving, coastal visits and ecotourism in the last decades (Kenny and Rees, 1994). This implies that biodiversity levels, at least in these areas, is sufficient to meet the needs of the existing users.

Data for fish catches and comments from the National Federation of Sea Anglers suggest concerns over the numbers and diversity of fish available for anglers (Roskilly, 2005). Seabird and marine mammal populations are generally increasing but in many cases should be regarded as fragile. Many recreational activities potentially damage the marine environment – angling, cruising and diving boats, for example, may damage fragile seabed habitats when anchoring, visitors to the shore may impact the system through trampling, litter and disturbance. Management therefore has to balance the need to protect the environment with the positive benefits of a larger proportion of the population engaging in nature-based recreation (Fletcher and Frid, 1997).

The major pressures on UK marine ecosystems

If the UK seabed is considered on a two by two nautical mile grid, 28 marine landscape classes can be recognized. Analysis of the distribution of the physical impacts of human activities on these landscape units shows that these were not homogeneous (Hall et al, 2007). Eight of the landscape types were subjected to very little pressure from human activities (less than 1 per cent of the area impacted). Impacts from activities such as construction of oil/gas rigs or wind farms occurred in many habitats although the total area affected was small. Overall, disturbance from bottom-towed fishing gears was the most extensive pressure. Siltation, from the plumes from aggregate extraction operations and spoil disposal, was the only other class of pressure to impact on more than 5 per cent of the area of any of the landscape types.

This section considers, for each of the five ecosystem goods and services categories being considered, the risk of collapse. This includes an assessment of the evidence of declines/deteriorations in the ecosystem components that deliver the 'goods and services' and the level of 'redundancy' in the system, in other words, if a component fails, is there another that would still be able to deliver the same goods/services?

The assessment is constrained by the some limitations; a lack of available data for all components covering the necessary areas and timescales. However, logic would suggest that a comprehensive assessment is not necessarily required, because if there is a risk of failure in any part of the system, then it needs to be addressed.

In the context of interdisciplinary consultation, it is easy to assume that the full range of ecosystem goods and services have been identified and quantified, and that all the ecosystem processes and functions underlying them are known and their interrelationships traced. It is self-evident that many subtle relationships exist among species and the environment, and yet we have hardly begun to understand the full complexity of any ecosystem, especially in relation to emergent properties. A key issue in marine environment research is that observation through monitoring does not necessarily convert to insight and understanding. Effects can be determined, but cause may only be inferred, assumed or simply not known. While the evidence base is added to through monitoring, the issue of circumstantial understanding impacts significantly on identifying or assessing the effectiveness of any conservation or remedial measures. Consequently, model scenarios based on our best understanding of marine ecosystem behaviour and response play a critical role in precautionary or adaptive management in which model output is validated using time series of data from monitoring stations or surveys. It is through an iterative process that models are validated and reconfigured to provide us with a better understanding of perhaps unseen or emergent phenomena that are not immediately apparent at the outset. Here, numerical modelling does not undermine the value of monitoring but adds additional weight to maintaining and even expanding our current data-gathering activities. In an economic climate of dwindling resources for survey and data acquisition, perhaps it is an essential component of future management activity that monitoring by the various regulatory and scientific bodies is more effectively coordinated.

The quality and diversity of the information used for this assessment of pressures on UK marine ecosystems implies a qualitative rather than a quantitative assessment. The quality of evidence and the risks are assessed using a standard terminology to aid comparisons. These are defined in Table 4.5.

Given the nature of the data and the still immature nature of the science in the area of biodiversity and ecological function, all conclusions drawn are tentative. Nor should it be assumed that stopping a damaging activity will necessarily cause the system to revert to its pre-impact state.

Timescales for changes

Given the high degree of natural spatial and temporal variability in marine ecosystems, long-term monitoring at relatively high intensities is required in order to understand variability patterns and to be able to comment meaningfully on their dynamics. It is also the case that many marine systems show a high degree of resilience, allowing recovery from disturbances (see Box 4.1).

Extreme perturbations can lead to shifts in the dynamics of the system, so-called phase shifts or alternative stable states (Frid et al, 2000). Here, potentially irreversible switching between contrasting ecosystem states, any of which may be considered resilient under given environmental conditions, variability and external stressors, may be considered as an unpredictable threshold-driven phenomenon that cannot be revealed by monitoring that is attuned to the detection of gradual, steady declines or recoveries (in other words, variation that is apparently normal), although Carpenter and Brock (2006) imply that increased variability precedes or at least forewarns of regime shifts between alternative stable states. Emergent phenomena such as alternative stable states may not even be predictable using currently available modelling approaches based on reductionist science. In this respect, monitoring or precaution cannot guard against switching between these states, and the potentially disastrous consequences for the delivery of ecosystem goods and services. One of the clearest examples of such a shift in marine systems has been in the Black Sea following the fishery collapse and subsequent invasion by a non-native species of jellyfish (Robinson and Frid, 2005).

The observed impact of introduced species and recovery of marine systems following impacts both show that marine ecosystems have the potential to undergo very large changes in short time frames. Observation and theory would suggest that, in many cases, species diverse systems do not show a gradual, steady, decline with increasing impact; rather systems may show little change with increasing pressure and then suddenly change dramatically. In the current context this implies that ecological functioning would initially be maintained but could suddenly collapse.

Given our limited knowledge of marine system dynamics, this argues for management regimes that involve a precautionary approach with a high degree of risk avoidance. Monitoring is not likely to detect that the system has reached the threshold until after it has passed it and key ecological functions are no longer available. This interpretation has been placed on the Canadian Grand and

---

Box 4.1 *Seabed recovery*

The sea floor off Northumberland was used for the dumping of ash from a power station. While this was occurring, an area of around 14km² was essentially without life as, almost every day, the dumped ash smothered the seabed. Following the end of waste dumping at sea in 1992, the seabed began a process of recovery and within nine months all areas had been colonized. A community similar to the background community was established within three years (Herrando-Pérez and Frid, 2001).

Studies following gravel extraction show considerable variability between sites, but generally periods of rapid improvement are followed by a prolonged phase of further continual improvement. Full recovery may take more than ten years, although if long-lived, slow-growing, species have been lost then complete recovery may never occur (Kenny and Rees, 1994, 1996; Boyd et al, 2005; Cooper et al, 2005).

Georges Banks cod collapses and the subsequent lack of recovery even after more than 15 years of complete fisheries closures (J. Rice Department of Fisheries and Oceans), personal communication.

While it is important to acknowledge the 'unknowns' in marine ecosystem science, one cannot argue against the growing consensus that it is time to act in order to protect marine biodiversity and to prevent further loss of marine ecosystem goods and services. Worm et al (2006), for example, identify worrying trends of collapse and extinction of species in coastal oceans over the past 1000 years, and global loss of species from large marine ecosystems over the last 50 years that amounts to a cumulative 65 per cent of collapsed taxa, increasing to as much as 80 per cent in species-poor ecosystems. This further illustrates the buffering role of species diversity on the resilience of ecosystem services. Although it is certain that our underpinning knowledge and evidence base increase with time and as new technologies and approaches come online (such as molecular systematics), this 'ideal' has to be balanced against the practical element of the extent of loss. Hence, the overriding need for immediate precautionary action.

Table 4.5 Definitions of standard terms used to define the quality of the evidence for changes in critical ecosystem components and the associated risk to delivery of ecosystem goods and services

Quality of the evidence			*Risk*
None	No evidence on which to form a judgement.	**Low**	Existing levels of change, or continuation of the observed trend, is unlikely to result in failure of the ecosystem to provide the goods/services or the scope for provision to be met from other parts of the ecosystem.
Some	Evidence of changes in the ecosystem components is restricted to a limited spatial area, confounded with other changes or largely inferred.	**Moderate**	Existing levels of change, or continuation of the observed trend, is likely to result in failure of the ecosystem to provide the current levels, or the full range of, goods/services. There might be some scope for provision to be partly met from other parts of the ecosystem.
Good	Clear evidence from a number of studies/areas that show deterioration in key ecosystem components.	**High**	Existing levels of change, or continuation of the observed trend is likely to result in failure of the ecosystem to provide the goods/services and there is little or no scope for provision to be met from other parts of the ecosystem.

Risk evaluation for atmospheric gas assimilation and climate regulation

There is good evidence from the continuous plankton recorder of changes in both the level and seasonal pattern of phytoplankton blooms around the UK (Reid et al, 1998). Changes appear to be driven more by climatic variations than nutrient inputs, but elevated nutrients may contribute both to the level of production and alterations in the species composition of blooms. These changes will influence the production of the atmospherically active DMS gas (Gabric et al, 2001) but calculations suggest that changes in *Emiliana huxleyi* and other algal blooms are unlikely to influence longer term sequestering of CO_2 (www. noc.soton.ac.uk/soes/staff/tt/ch/biogeochemistry.html). In part, this is due to species replacements occurring, such that the total drawdown of CO_2 remains unchanged. This conclusion is tentative and some studies contradict it. Furthermore ecological theory and studies in enclosed bodies, such as freshwater lakes, suggest that it is likely that once a threshold is past, species replacements will no longer compensate for loss. There are no data to allow prediction of this threshold for collapse.

Conclusion

There is good evidence of deterioration in the ecosystem components that support the ecosystem services that regulate the atmosphere and climate. There is a low-moderate risk that these changes could lead to undesirable consequences for the atmosphere and climate.

Risk evaluation for nutrient recycling capacity

Although there is evidence that the organisms associated with nutrient recycling have been affected by human activities over the last century, there is little evidence to suggest that the total rate of these processes has been altered (Frid and Paramor, 2006).

Conclusion

There is some evidence of deterioration in the ecosystem components that support the ecosystem services that provide nutrient recycling. There is a low risk that these changes could lead to undesirable consequences for the recycling of nutrients.

Risk evaluation for loss of waste assimilation capacity

Elevated nutrients, from agricultural run-off and sewage treatment, probably contribute both to the altered level and pattern of phytoplankton production and the altered species composition of blooms. These changes could lead to altered food web dynamics and hence altered drawdown of nutrients and hence altered waste assimilation capacity for organic wastes.

Improvements in waste disposal regulation in recent decades has decreased pressure on the system's waste assimilation capacity and, except in localized areas, there appears to be low risk of loss of this function if current trends in waste disposal continue.

Conclusion

There is good evidence of deterioration in the ecosystem components that support the ecosystem services that provide waste assimilation. There is a low (moderate for nutrient-containing wastes) risk that these changes could lead to undesirable consequences for ability of the system's waste assimilation capacity.

Risk evaluation for habitat functions

There is good evidence that habitats have been affected by human activities. Some habitats are very sensitive to disturbance and may be permanently destroyed by a single impact (for example, cold water coral reefs), whilst other habitat types may be more tolerant to disturbance (for example, sands). There is evidence that long-term and high frequency activities, such as fishing, have affected their ability to support 'natural' communities and there exists the very real possibility that major, and ecologically important, changes occurred prior to any scientific studies. This is suggested by studies using data from the early 20th century (Frid et al, 2000; Rumohr and Kujawski, 2000) and the fossil record (Robinson and Frid, 2005).

There is good evidence of losses to biogenic habitat types such as seagrass beds (UK BAP, 1995; Davison and Hughes, 1998; Langston et al, 2006), shellfish reefs (oyster and mussel beds) (Riesen and Reise, 1982; UK BAP, 1999), maerl (Hall-Spencer and Moore, 2000; Barbera et al, 2003), worm reefs (*Sabellaria alveolata)* (Riesen and Reise, 1982; Vorberg, 2000) and cold water coral (*Lophelia*) reefs (Mortensen et al, 2001; Fosså et al, 2002; Roberts et al, 2003) within the past century as a result of human activities. Loss or damage to these habitat types will result in a reduction of the high levels of biological diversity, and the functions these communities provide.

There is evidence to suggest that damage to some of the more common, and more resilient, habitat types, such as sands and gravels, has resulted following activities such as aggregate extraction. Some studies have observed clear differences in benthic communities for several years after the cessation of dredging (Boyd and Rees, 2003; Boyd et al, 2005). This may affect the natural functioning of the area.

Conclusion

There is good evidence of deterioration in the ecosystem components that support the ecosystem services that provide societal value. There is a high risk that these changes could lead to undesirable consequences for the delivery of habitat functions.

Risk evaluation for food provision

There is good evidence of changes in fish populations around the UK. These cover both fish species targeted by the fisheries but also the non-target members of the fish community (Heesen and Daan, 1996; Pope and Macer, 1996). Some of these changes can be attributed to variations/changes in climate and hydrography, although the single greatest influence is fishing. Given the economic value of fisheries (PMSU, 2004) and role of fish in the diet of many species of high public interest (seabirds, marine mammals), the decline of the fish component of the ecosystem is likely to have wide-ranging negative effects both on food supply and on biodiversity, and particularly the components most valued by society.

In the North Sea alone, eight species of demersal fish consume around 28 million tonnes of benthic food (worms, clams, brittlestars, sea urchins, crabs, shrimps, and so on) each year (Frid et al, 1999). As fish species vary in their ability to capture different prey, changes in the amount of benthic food and the composition of the benthic community may potentially impact on the food resource of these fish.

There is clear evidence for a series of changes in plankton in UK waters. These include declines in some key taxa (for example, *Calanus finmarchicus*), but also altered levels and patterns of productivity, and changes in species composition (some species are doing better, some worse). Given that plankton is the base of the marine food chains, these changes are potentially highly destabilizing.

The strong link between shelf benthic communities and the productivity of the overlying water means changes in the amount and probably the timing of plankton production will have implications for benthic communities. This will, in turn, alter the food supply for demersal fish.

There is some evidence that changes in plankton has affected the growth and survivorship of larval fish and may be contributing to the decline in some fish populations.

Conclusion

There is good evidence of deterioration in the ecosystem components that support the ecosystem services that support food provision. There is a high risk that these changes could lead to further undesirable consequences for the supply of food from the sea.

Risk evaluation for biodiversity in support of societal values

Seabirds have a high public profile and are economically important through eco-tourism and recreational use of biological diversity. They are also protected under UK and European law and numerous international conventions.

While there is good evidence of a decline in the occurrence of many species of seabird when compared with data from previous centuries, there is also clear evidence of recovery in many species following the introduction of protective measures (Defra, 2005). Seabirds fulfil a wide range of ecological roles but do not appear to be the critical group for the delivery of any particular ecosystem service other than biologically diverse and naturally functioning systems. These species tend to be at or near the apex of the food web and their loss represents a truncation of the natural food web.

Marine mammals and reptiles have a high public profile and are subject to protection under UK and European law and various international conventions. Marine mammals are often the key resource in marine ecotourism ventures.

There is good evidence for declines in the occurrence of some marine mammals and reptiles but also evidence of recovery in some groups, such as seals. These species are not the key for the delivery of any ecosystem services other than biologically diverse and naturally functioning systems. These species tend to be at or near the apex of the food web and their loss represents a truncation of the natural food web.

There is good evidence of changes in fish populations around the UK. These changes cover both fish species targeted by the fisheries and recreational sea anglers, but also the non-target members of the fish community (Heesen and

Daan, 1996; Pope and Macer, 1996). Some of these changes can be attributed to variations/changes in climate and hydrography. However, the single greatest influence is fishing. Given the economic value of fisheries (PMSU, 2004) and role of fish in the diet of many species of high public interest (seabirds, marine mammals), a decline of the fish component of the ecosystem is likely to have wide-ranging negative effects both on food supply and on biodiversity, and particularly the components most valued by society.

To date there is no clear evidence of negative effects on birds and shellfisheries of increased frequency of blooms of toxic and nuisance algae.

Conclusion

There is good evidence of deterioration in the ecosystem components that support the ecosystem services that provide societal value. There is a high risk that these changes could lead to undesirable consequences for recreational activities supported by marine biodiversity.

Marine ecosystem services and marine spatial planning

This brief assessment has described the basis of an ecosystem services approach and applied this to the UK marine environment to highlight both the evidence base for assessing the health of these ecosystems and outlined the ongoing risks to their wider ecological functions. This appears to confirm the findings of other assessments (for example, MEA, 2005; Worm et al, 2006) that have highlighted how the loss of biodiversity in marine environments is now impairing their capacity to support human life through the functions described in this paper.

Table 4.6 Summary of the analysis of the evidence that biological systems that deliver ecological services have been compromised and the risk to the continued delivery of the service

Ecological services	Quality of evidence of a deterioration in ecosystem providers	Risk to delivery of ecological service
Gas and climate regulation	Good	Low-moderate
Nutrient cycling	Some	Low
Waste treatment	Good	Low (moderate for nutrient containing wastes)
Habitat functions	Good	High
Food and material provision	Good	High
Biodiversity in support of societal values	Good	High

It is for this reason that there has been increasing demands for a more coherent and comprehensive system of marine management, to which the UK's Marine and Coastal Access Act 2009 is one response. This has ultimately evolved, through a growing evidence base and exposure to the realpolitik of the policy process, from the new framework for environmental management initiated by the 1992 United Nations Convention on Biological Diversity (CBD). One key principle of this was the recognition that people are part of the ecosystem and that, while use of ecological goods and services inevitably lead to impacts on the system, the scale of these must be balanced against derived benefits and underpinned by sustainability criteria to safeguard options of future generations. Furthermore there was a requirement for environmental management to follow an 'ecosystem approach', which requires management to give consideration to the full diversity of ecosystem components and to recognize the cumulative effects of the range of human pressures on the system. One of the main challenges to such integration in the past has been the lack of a common currency to understand ecosystem impacts from various sources. Recent advances in our understanding of the spatial structure of marine ecosystems, the development of novel techniques to link biological entities with the delivery of ecological functions and the emergence of spatial maps of human impacts mean that a common, spatial, framework now exists. Indeed, as this chapter has shown, marine environments differ geographically and temporally in terms of how they contribute to natural and human systems and in their sensitivity to human activity, providing an essential evidence base for management regimes that can maximize the human utility of current marine ecosystems and safeguard their integrity for future generations.

In the earlier part of this chapter we considered elements of ecosystem health and resilience, and that the delivery of ecosystem goods and services is dependent upon good ecological status. However, healthy ecosystems are not necessarily those that maximize the delivery of ecosystem goods and services for human health and well-being. In adopting an ecosystem approach where management priorities are socially defined, there is a real danger that ecosystem function and resilience are severely compromised – especially if marine management goals are set at the scale of local or regional sea. A full consideration of marine ecosystem goods and services requires an understanding of their importance at a variety of time and length scales, and especially the global importance of the ocean in climate regulation through both biogeochemical and physical interaction with the atmosphere. Does this global climate regulation service have a value to humans that can be made tangible at the decision-making level? Societal and/or economic choice in relation to the delivery of marine ecosystem goods and services is much more likely to be linked to particular seas and their surrounding nation states, especially as economies become less global and more local. Cities, which are the primary consumers of food, are likely to require shorter supply lines as oil prices continue to rise, so that local food production will probably be economically important. In this sense, global supporting and regulating ecosystem services may suffer at the expense of provisioning services expressed at the local scale. In addition societal/economic importance is time-dependent as priorities change or become changed. By their nature, ecosystems and the services they provide to human societies are taken for granted until some failure to deliver highlights a problem. One aspect of the current concern about ecosystem services is that they are almost universally threatened. Indeed, we are experiencing what is believed to be the Earth's greatest ever mass extinction (Harte et al, 2004). This means that the less threatened services, such as gas and climate regulation, nutrient cycling and waste treatment, may eventually become at high risk of deterioration.

The value of goods and services linked to conservation may also be overlooked in defining management priorities, as they have often been overlooked in the past. Balmford et al (2002) considered the value of retaining the undisturbed condition for a variety of ecosystems against being converted for economic development, as illustrated by the difference in benefit flows of goods and services. In all cases considered, the loss of non-marketed services was greater than the marginal benefits of conversion. It is, therefore, demonstrated that for all ecosystem goods and services considered, conservation enhances human well-being above that of conversion under the auspices of economic development.

A key question in the management of marine ecosystem goods and services is the development of management practice. If remediation is to be undertaken, how is it possible to identify reference baseline states – especially in the context of incomplete understanding and data, inherent ecosystem variability and response, and underlying trajectories linked to climate change and human impact? Similarly, it may be argued that no pristine marine ecosystems exist as analogues for remediation targets. In addition, how do we go about determining the spatial and temporal limits of management practices when the connectivity between ecosystems is so critical? In coastal waters especially, what happens on the seabed and within the water column is just as much, if not more so, determined by activity on land as it is in the sea. Ecosystems are also in a constant state of flux – driven all the more by climate change. Climate change makes extinctions in marine areas a logical consequence – even for presently healthy, resilient ecosystems – as the limits of, and variability in, environmental conditions alter through time. This dictates management for change rather than conserving/ preserving what we have, making management targets less definable. Marine Reserves, Marine Protected Areas (MPAs) and Marine Conservation Zones (MCZs) perhaps deserve more attention as these are perhaps the cornerstone of management policy and practice for the time being. According to Worm et al

(2006) MPAs and Marine Reserves offer the potential to: reverse decline in biodiversity on local/regional scales; increase fisheries productivity; increase recovery after natural disturbances; deliver reduced community variability; and provide increased tourist revenue. In this respect, the benefits of implementation for marine ecosystem goods and services appear considerable. However, there are practical issues linked to the scale of protected areas. Depending on the ecosystem component (or components) that are the focus of protection, the extent of the protected area must be attuned to their function. The role of the oceans in regulating climate indirectly illustrates the importance of ecosystem connectedness and corridors or migratory pathways. Irrespective of where and how, and the mechanism by which the success of protected areas is measured, the precautionary approach is clearly of utmost importance.

In many countries, including the UK, a central mechanism for marine management is now Marine Spatial Planning (see, for example, Douvere, 2008), established in the UK through the 2009 Act. This represents the mechanism by which ecosystem-based management is implemented and mediated with other government objectives, such as economic growth and human welfare, through deliberative, integrated and forward-looking assessments of future human needs and environmental sensitivities. This inevitably leads to trade-offs between ecological impacts and the interests of other, human-based stakeholders, such as the fishing or aggregates sectors. In the absence of a comprehensive system of MSP, such decisions tended to be made in an ad hoc and fragmented approach, with cumulative impacts being poorly monitored and assimilated into decision-making. The establishment of a system of MSP offers the potential to anticipate the ecosystem-wide impacts of current and future development and designate, in temporal and spatial terms, appropriate sites for development, or indeed the biological process most critical to different ecosystem services – for example through marine protection areas (Ehler, 2008). Spatial planning has long faced problems in assimilating the heterogeneous data, values and forms of knowledge needed to make the best informed decisions (for example, Moroni, 2006), which has involved attempts at developing common assessment frameworks, such as planning balance sheets, sustainability assessments and the natural capital approach. This is no less appropriate to the marine environment and the emerging discipline of MSP, which faces particular difficulties as less is known about marine ecosystems and how they may respond to human induced stress. While it is inevitable that this will stimulate the emergence of new assessment tools and evaluation frameworks, the ecosystem services approach is a robust basis for identifying critical elements of marine ecosystems and ensuring that such values are fully reflected in marine management decisions.

Conclusions

In this chapter we have acknowledged the high level of specialist knowledge required to see through the ambiguity in ecosystem concept terminology, yet we recognize ecosystem goods and services as a key tool for transdisciplinary discourse that can transcend disciplinary specialism.

It is also recognized that is it time to acknowledge the unknowns in our understanding of marine ecosystems but that it is time to act to address loss of biodiversity through precautionary and/or adaptive management.

Acknowledgements

The thoughts expressed in this chapter have developed over many years and through working with many colleagues in the UK and internationally. This chapter has developed from work carried out with Odette Paramor as part of the evidence produced by Defra in support of the Marine Bill. We gratefully acknowledge Defra's support and Odette's thoughts and ideas. The opinions in this paper are those of the authors and do not represent those of University of Liverpool, Defra or any other individual. The authors also take responsibility for the errors and omissions.

Note

1   Constanza et al (1997) considered categories 4–7 and 10 (water regulation, water supply, erosion control and sediment retention, soil formation, pollination) as not being relevant to marine ecosystems.

References

Agardy, T. (2003) 'An environmentalist's perspective on responsible fisheries: The need for holistic approaches', in M. Sinclair and G. Valdimarsson (eds) *Responsible Fisheries in the Marine Ecosystem*, FAO, Rome and CABI Publishing, Wallingford, pp65–85

Aller, R. C. (1982) 'The effects of macrobenthos on chemical properties of marine sediment and overlying water', in P. L. McCall and M. J. S. Tevesz (eds) *Animal-sediment Relations*, Plenum Press, New York, pp53–102

Aller, R. C. (1988) 'Benthic fauna and biogeochemical processes in marine sediments: The role of burrow structures', in T. H. Blackburn and J. Sorensen (eds) *Nitrogen*

Recycling in Coastal Marine Environments, John Wiley and Sons Ltd, London, pp301–338

Arkema, K., Abramson, S. and Dewsbury, B. (2008) 'Marine ecosystem-based management: From characterization to implementation', *Frontiers in Ecology and the Environment*, vol 4, no 10, pp525–532

Balch, W. M., Holligan, P. M. and Kilpatrick, K. A. (1992) 'Calcification, photosynthesis and growth of the bloom-forming coccolithophore, *Emiliania huxleyi*', *Continental Shelf Research*, vol 12, pp1353–1374

Balmford, A., Bruner, A., Cooper, P., Costanza, R., Farber, S., Green, R. E., Jenkins, M., Jefferiss, P., Jessamy, V., Madden, J., Munro, K., Myers, N., Naeem, S., Paavola, J., Rayment, M., Rosendo, S., Roughgarden, J., Trumper, K. and Turner, R. K. (2002) 'Economic reasons for conserving wild nature', *Science*, vol 297, pp950–953

Barbera, C., Bordehore, C., Borg, J. A., Glémarec, M., Grall, J., Hall-Spencer, J. M., de la Huz, C., Lanfranco, E., Lastra, M., Moore, P. G., Mora, J., Pita, M. E., Ramos-Esplá, A. A., Rizzo, M., Sánchez-Mata, A., Seva, A., Schembri, P. J. and Valle, C. (2003) 'Conservation and management of northeast Atlantic and Mediterranean maerl beds', *Aquatic Conservation: Marine and Freshwater Ecosystems*, vol 13, no 1, ppS65–S76

Beaumont, N. J. and Tinch, R. (2003) *Goods and Services Related to the Marine Benthic Environment*, CSERGE Working Paper ECM 03-14, p12

Bengtsson, J. (1998) 'Which species? What kind of diversity? Which ecosystem function? Some problems in studies of relations between biodiversity and ecosystem function', *Applied Soil Ecology*, vol 10, pp191–199

Bigg, G. R., Jickells, T. D., Liss, P. S. and Osborn, T. J. (2003) 'The role of the oceans in climate', *International Journal of Climatology*, vol 23, no 10, pp1127–1159

Biles, C. L., Paterson, D. M., Ford, R. B., Solan, M. and Raffaelli, D. G. (2002) 'Bioturbation, ecosystem functioning and community structure', *Hydrology and Earth System Sciences*, vol 6, no 6, pp999–1005

Blackford, J. C. (1997) 'An analysis of benthic biological dynamics in a North Sea ecosystem model', *Journal of Sea Research*, vol 38, pp213–230

Bolam, S. G., Rees, H. L., Somerfield, P., Smith, R., Clarke, K. R., Warwick, R. M., Atkins, M. and Garnacho, E. (2006) 'Ecological consequences of dredged material disposal in the marine environment: A holistic assessment of activities around the England and Wales coastline', *Marine Pollution Bulletin*, vol 52, no 4, pp415–426

Bolger, T. (2001) 'The functional value of species biodiversity: A review', *Biology and Environment: Proceedings of the Royal Irish Academy*, vol 101B, no 3, pp199–234

Boyd, S. E. and Rees, H. L. (2003) 'An examination of the spatial scale of impact on the marine benthos arising from marine aggregate extraction in the central English Channel', *Estuarine, Coastal and Shelf Science*, vol 57, pp1–16

Boyd, S. E., Limpenny, D. S., Rees, H. L. and Cooper, K. M. (2005) 'The effects of marine sand and gravel extraction on the macrobenthos at a commercial dredging site (results 6 years post-dredging)', *ICES Journal of Marine Science*, vol 62, no 2, pp145–162

Bremner, J., Rogers, S. I. and Frid, C. L. J. (2006) 'Matching biological traits to environmental conditions in marine benthic ecosystems', *Journal of Marine Systems*, vol 60, pp302–316

Carpenter, S. R. and Brock, W. A. (2006) 'Rising variance: A leading indicator of ecological transition', *Ecology Letters*, vol 9, no 3, pp311–318

Clark, J., Burgess, J. and Harrison, C. M. (2000) '"I struggled with this money business": Respondents' perspectives on contingent valuation', *Ecological Economics*, vol 33, no 1, pp45–52

Clark, R. B., Frid, C. L. J. and Attrill, M. (1997) *Marine Pollution*, 4th edition, Clarendon Press, Oxford, ppxii and 161

Connor, D. W., Gilliland, P. M., Golding, N., Robinson, P., Todd, D. and Verling E. (2006) *UKSeaMap: The Mapping of Seabed and Water Column Features of UK Seas*, Joint Nature Conservation Committee, Peterborough

Cooper, K. M., Eggleton, J. D., Vize, S. J., Vanstaen, K., Smith, R., Boyd, S. E., Ware, S., Morris, C. D., Curtis, M., Limpenny, D. S. and Meadows, W. J. (2005) 'Assessment of the rehabilitation of the seabed following marine aggregate dredging: Part II', in *Scientific Series, Technical Report 130*, Cefas, Lowestoft

Costanza, R., D'Arge, R., de Groot, R., Farber, S., Grasso, M., Hannon, B., Limburg, K., Naeem, S., O'Neill, R. V., Paruelo, J., Raskin, R. G., Sutton, P. and van den Belt, M. (1997) 'The value of the world's ecosystem services and natural capital', *Nature*, vol 387, no 6630, pp253–260

Davison, D. M. and Hughes, D. J. (1998) *Zostera Biotopes (volume I): An Overview of Dynamics and Sensitivity Characteristics for Conservation Management of Marine SACs*, Scottish Association for Marine Science (UK Marine SACs Project)

Defra (Department for Environment, Food and Rural Affairs) (2002) *Safeguarding Our Seas: A Strategy for the Conservation and Sustainable Development of our Marine Environment*, Defra, London

Defra (2005) *Charting Progress: An Integrated Assessment of the State of the UK Seas*, Defra, London

Diaz, S. and Cabido, M. (2001) 'Vive la difference: Plant functional diversity matters to ecosystem processes', *Trends in Ecology and Evolution*, vol 16, no 11, pp646–655

Douvere, F. (2008) 'The importance of marine spatial planning in advancing ecosystem-based sea use management', *Marine Policy*, vol 32, no 5, pp762–771

Drew Associates Ltd (2003) *Research into the Economic Contribution of Sea Angling*, Defra, London, p82

EC (European Commission) (1979) *Council Directive of 2 April 1979 on the Conservation of Wild Birds*, 79/409/EEC, EC

EC (1991) *Council Directive of 21 May 1991 Concerning Urban Waste-water Treatment*, 91/271/EEC, EC

EC (2002) *Proposal for a Directive of the European Parliament and of the Council Concerning the Quality of Bathing Water*, EC

EC (2007) *An Integrated Maritime Policy for the European Union*, EC, p575

Edwards, M., Richardson, A. J., Batten, S. and John, A. W. G. (2004) 'Ecological status report: Results from the CPR survey 2002/2003', in *Technical Report, No 1*, SAHFOS, pp1–8

Ehler, C. (2008) 'Conclusions: Benefits, lessons learned, and future challenges of marine spatial planning', *Marine Policy*, vol 32, no 5, pp840–843

Fletcher, H. and Frid, C. L. J. (1997) 'Impact and management of visitor pressure on rocky intertidal algal communities', *Aquatic Conservation-Marine and Freshwater Ecosystems*, vol 7, no 1, pp287–297

Fosså, J. H., Mortensen, P. B. and Furevik, D. M. (2002) 'The deep-water coral *Lophelia pertusa* in Norwegian waters: Distribution and fishery impacts', *Hydrobiologica*, vol 471, pp1–12

Fowler, D., Pilegaard, K., Sutton, M. A., Ambus, P., Raivonen, M., Duyzer, J., Simpson, D., Fagerli, H., Fuzzi, S., Schjoerring, J. K., Grance, C., Neftel, A., Isaksen, I. S. A., Laj, P., Maione, M., Monks, P. S., Burkhardt, J., Daemmgen, U., Neirynck, J., Personne, E. et al (2009) 'Atmospheric composition change: Ecosystems-atmosphere interaction', *Atmospheric Environment*, vol 43, no 33, pp5193–5267

Frid, C. L. J. and Paramor, O. A. L. (2006) *Marine Biodiversity and the Rationale for Intervention*, report to Defra from the School of Biological Sciences, University of Liverpool

Frid, C. L. J. Hammer, C., Law, R. J., Loeng, H., Pawlak, J. E., Reid, P. C. and Tasker, M. (2003) *Environmental Status of the European Seas*, ICES, Copenhagen

Frid, C. L. J., Hansson, S., Ragnarsson, S. A., Rijnsdorp, A. and Steingrimsson, S. A. (1999) 'Changing levels of predation on benthos as a result of exploitation of fish populations', *Ambio*, vol 28, no 7, pp578–582

Frid, C. L. J., Harwood, K. G., Hall, S. J. and Hall, J. A. (2000) 'Long-term changes in the benthic communities on North Sea fishing grounds', *ICES Journal of Marine Science*, vol 57, no 5, pp1303–1309

Frid, C. L. J., Paramor, O. A. L. and Scott, C.L. (2006) 'Ecosystem-based management of fisheries: Is science limiting?', *ICES Journal of Marine Science*, vol 63, pp1567–1572

Fritz, J. S. (2010) 'Towards a "new form of governance" in science-policy relations in the European Maritime Policy', *Marine Policy*, vol 34, no 1, pp1–6

Gabric, A., Gregg, W., Najjar, R., Erickson, D. and Matrai, P. (2001) 'Modelling the biogeochemical cycle of dimethylsulfide in the upper ocean: A review', *Chemosphere – Global Change*, vol 3, no 4, pp377–392

Galay Burgos, M. and Rainbow, P. S. (2001) 'Availability of cadmium and zinc from sewage sludge to the flounder, *Platichythys flesus*, via a marine food chain', *Marine Environmental Research*, vol 51, pp417–439

Gooday, A. J. (2002) 'Biological responses to seasonally varying fluxes of organic matter to the ocean floor: A review', *Journal of Oceanography*, vol 58, pp305–332

Gowen, R. J., Mills, D. K., Timmer, M. and Nedwell, D. B. (2000) 'Production and its fate in two coastal regions of the Irish Sea: The influence of anthropogenic nutrients', *Marine Ecology Progress Series*, vol 208, pp51–64

Hall, K., Winrow-Giffin, A., Paramor, O. A. L., Robinson, L. A. and Frid, C. L. J. (2007) *Mapping the Sensitivity of Benthic Habitats to Fishing in Welsh Waters– Development of a Protocol*, CCW (Countryside Council for Wales), Bangor

Hall-Spencer, J. M. and Moore, P. G. (2000) 'Impact of scallop dredging on maerl grounds', in M. J. Kaiser and S. J. de Groot (eds) *Effects of Fishing on Non-target Species and Habitats: Biological Conservation and Socio-economic Issues*, Blackwell Science, Oxford, pp105–117

Harte, J., Ostling, A., Green J. L. and Kinzig, A. (2004) 'Biodiversity conservation: Climate change and extinction risk,' *Nature*, vol 430, no 6995, doi: 10.1038/nature02718

Hecky, R. E. and Kilham, P. (1988) 'Nutrient limitation of phytoplankton in freshwater and marine environments: A review of recent evidence on the effects of enrichment', *Limnology and Oceanography*, vol 33, no 4, pp796–822

Heesen, H. J. L. and Daan, N. (1996) 'Long term trends in non-target North Sea fish species', *ICES Journal of Marine Science*, vol 53, pp1063–1078

Herrando-Pérez, S. and Frid, C. L. J. (2001) 'Recovery patterns of macrobenthos and sediment at a closed fly-ash dumpsite', *Sarsia*, vol 86, pp389–400

Horowitz, A. J. and Elrick, K. A. (1987) 'The relation of stream sediment surface area, grain size and composition to trace element chemistry', *Applied Geochemistry*, vol 2, pp437–451

Howe, R. L., Rees, A. P. and Widdicombe, S. (2004) 'The impact of two species of bioturbating shrimp (*Callianassa subterranea* and *Upogebia deltaura*) on sediment denitrification', *Journal of the Marine Biological Association of the United Kingdom*, vol 84, no 3, pp629–632

Hughes, K. A. and Thompson, A. (2004) 'Distribution of sewage pollution around a maritime Antarctic research station indicated by faecal coliforms, *Clostridium perfringens*, and faecal sterol markers', *Environmental Pollution*, vol 127, pp315–321

ICES (International Council for the Exploration of the Sea) (2004) 'Report of the working group on the assessment of demersal stocks in the North Sea and Skagerrak', in *ICES Council Meet Pap 2004 ACFM:07*, pp1–555

ICES (2005) *Report of the Working Group on Ecosystem Effects of Fishing Activities (WGECO)*, ICES, Copenhagen

ICES (2006) *Report of the Working Group on Ecosystem Effects of Fishing Activities (ACE:05)*, ICES, Copenhagen

Jackson, D., Lambers, B. and Gray, J. (2000) 'Radiation doses to members of the public near to Sellafield, Cumbria, from liquid discharges 1952–1998', *Journal of Radioecological Protection*, vol 20, pp139–167

Kenny, A. J. and Rees, H. L. (1994) 'The effects of marine gravel extraction on the macrobenthos: Early post dredging recolonization', *Marine Pollution Bulletin*, vol 28, no 7, pp442–447

Kenny, A. J. and Rees, H. L. (1996) 'The effects of marine gravel extraction on the macrobenthos: Results two years post-dredging', *Marine Pollution Bulletin*, vol 32, nos 8 9, pp615–622

Langston, W., Chesman, B., Burt, G., Taylor, M., Covey, R., Cunningham, N., Jonas, P. and Hawkins, S. (2006) 'Characterisation of the European Marine Sites in south west England: The Fal and Helford candidate Special Areas of Conservation (cSAC)', *Hydrobiologia*, vol 555, pp321–333

Leah, R. T., Evans, S. J., Johnson, M. S. and Collings, S. (1991) 'Spatial patterns in accumulation of mercury by fish from the NE Irish Sea', *Marine Pollution Bulletin*, vol 22, no 4, pp172–175

Liss, P. S. (2002) 'Biogeochemical connections between the atmosphere and the ocean', *International Geophysics*, vol 83, pp249–258

Marine Fisheries Agency (2005) *United Kingdom Sea Fisheries Statistics 2004*, National Statistics/Defra, London

Marine Fisheries Agency (2008) *United Kingdom Sea Fisheries Statistics 2008*, www. marinemanagement.org.uk/fisheries/statistics/documents/ukseafish/2008/final.pdf

Matthiessen, P. and Law, R. J. (2002) 'Contaminants and their effects on estuarine and coastal organisms in the United Kingdom in the late twentieth century', *Environmental Pollution*, vol 120, pp739–757

MEA (Millennium Ecosystem Assessment) (2005) *Ecosystems and Human Well-being: Synthesis*, Island Press, Washington, DC

Moroni, S. (ed) (2006) *Evaluation in Planning: Evolution and Prospects*, Ashgate, Aldershot

Mortensen, P. B., Hovland, T., Fosså, J. H. and Furevik, D. M. (2001) 'Distribution, abundance and size of *Lophelia pertusa* coral reefs in mid-Norway in relation to seabed characteristics', *Journal of the Marine Biological Association of the UK*, vol 81, no 4, pp581–597

Naeem, S. and Wright, J. P. (2003) 'Disentangling biodiversity effects on ecosystem functioning: Deriving solutions to a seemingly insurmountable problem', *Ecology Letters*, vol 6, pp567–579

Naeem, S., Loreau, M. and Inchausti, P. (2004) 'Biodiversity and ecosystem functioning: The emergence of a synthetic ecological framework', in M. Loreau, S. Naeem, and P. Inchausti (eds) *Biodiversity and Ecosystem Functioning*, Oxford University Press, Oxford, pp3–11

NMP (National Monitoring Programme) (1998) *National Monitoring Programme: Survey of the Quality of UK Coastal Waters*, Marine Pollution Monitoring Management Group, Aberdeen

Nixon, S. W. (1981) 'Remineralisation and nutrient recycling in coastal marine ecosystems', in B. J. Neilson and L. E. Cronin (eds) *Estuaries and Nutrients*, Humana Press, Clifton NJ, pp111–138

Paramor, O. A. L., Scott, C. L. and Frid, C. L. J. (eds) (2004) *European Fisheries Ecosystem Plan: Producing a Fisheries Ecosystem Plan*, University of Newcastle upon Tyne

Percival, P., Frid, C. L. J. and Upstill-Goddard, R. (2005) 'The impact of trawling on benthic nutrient dynamics in the North Sea: Implications of laboratory experiments', *American Fisheries Society Symposium*, vol 41, pp491–501

Perez, M., Usero, J. and Gracia, I. (1991) 'Trace metals in sediments from the Ria de Huelva', *Toxicological and Environmental Chemistry*, vols 31–32, pp275–283

Phillips, D. J. H. and Rainbow, P. S. (1989) 'Strategies of trace metal sequestration in aquatic organisms', *Marine Environmental Research*, vol 28, pp207–210

Pilskaln, C. H., Churchill, J. H. and Mayer, L. M. (1998) 'Resuspension of sediment by bottom trawling in the Gulf of Maine and potential geochemical consequences', *Conservation Biology*, vol 12, no 6, pp1223–1229

Plasman, I. C. (2008) 'Implementing marine spatial planning: A policy perspective', *Marine Policy*, vol 32, no 5, pp811–815

PMSU (Prime Minister's Strategy Unit) (2004) *Net Benefits: A Sustainable and Profitable Future for UK Fishing*, PMSU, Cabinet Office, London

Pope, J. G. and Macer, C. T. (1996) 'An evaluation of the stock structure of North Sea cod, haddock and whiting since 1920, together with a consideration of the impacts of fisheries and predation effects on their biomass and recruitment', *ICES Journal of Marine Science*, vol 53, pp1157–1169

Prince, R. C. (1993) 'Petroleum spill bioremediation in marine environments', *Critical Reviews in Microbiology*, vol 19, no 4, pp217–242

Pugh, D. and Skinner, L. (2002) *A New Analysis of Marine-related Activities in the UK Economy with Supporting Science and Technology*, IACMST Information Document No 10, Inter-Agency Committee on Marine Science and Technology, Southampton

Rainbow, P. S., Geffard, A., Jeantet, A. Y., Smith, B. D., Amiard, J. C. and Amiard-Triquet, C. (2004) 'Enhanced food-chain transfer of copper from a diet of copper-tolerant estuarine worms', *Marine Ecology Progress Series*, vol 271, pp183–191

Rees, J. G., Ridgway, J., Knox, R., Wiggans, G. and Breward, N. (1998) 'Sediment-borne contaminants in rivers discharging into the Humber Estuary', *Marine Pollution Bulletin*, vol 31, nos 3–7, pp316 329

Reid, P. C., Edwards, M., Hunt, H. G. and Warner, A. J. (1998) 'Phytoplankton change in the North Atlantic', *Nature,* vol 391, no 6667, p546

Reid, P. C., Fischer, A. C., Lewis-Brown, E., Meredith, M. P., Sparrow, M., Andersson, A. J., Antia, A., Bates, N. R., Bathmann, U. et al, (2009) 'Impacts of the oceans on climate change', *Advances in Marine Biology*, vol 56, pp1–150

Riesen, W. and Reise, K. (1982) 'Macrobenthos of the subtidal Wadden Sea: Revisited after 55 years', *Helgoländer Meeresunters*, vol 35, pp409–423

Roberts, J. M., Long, D., Wilson, J. B., Mortensen, P. B. and Gage, J. D. (2003) 'The cold-water coral *Lophelia pertusa* (Scleractinia) and enigmatic seabed mounds along the north-east Atlantic margin: Are they related?', *Marine Pollution Bulletin*, vol 46, no 1, pp7–10

Robinson, L. A. and Frid, C. L. J. (2005) 'Extrapolating extinctions and extirpations: Searching for a pre-fishing state of the benthos', *American Fisheries Society Symposium*, vol 41, pp619–628

Roskilly, L. (2005) *Marine Protected Areas and Recreational Sea Angling*, National Federation of Sea Anglers Conservation Group

Rumohr, H. and Kujawski, T. (2000) 'The impact of trawl fishery on the epifauna of the southern North Sea', *ICES Journal of Marine Science*, vol 57, pp1389–1394

Sagoff, M. (1998) 'Some problems with environmental economics', *Environmental Ethics*, vol 10, no 1, pp55–74

Sly, P. G. (1989) 'Sediment dispersion: Part 2, characterisation by size of sand fraction and percent mud', *Hydrobiologia*, vols 176–177, no 1, pp111–124

Somerville, H. J., Bennett, D., Davenport, J. N., Holt, M. S., Lynes, A., Mahieu, A., McCourt, B., Parker, J. G., Stephenson, R. R., Watkinson, R. J. and Wilkinson, T. G. (1987) 'Environmental effect of produced water from North Sea oil operations', *Marine Pollution Bulletin*, vol 18, no 10, pp549–558

Stern, N. (2006) *The Economics of Climate Change*, HM Treasury, London

Swannell, R. P. J., Lee, K. and McDonagh, M. (1996) 'Field evaluations of marine oil spill bioremediation', *Microbiology and Molecular Biology Reviews*, vol 60, no 2, pp342–365

Tarpgaard, E., Mogensen, M., Grønkaer, P. and Carl, J. (2005) 'Using short term growth of enclosed Ø-group European flounder, *Platichthys flesus*, to assess habitat quality in a Danish Bay', *Journal of Applied Ichthyology*, vol 21, no 1, pp53–63

Trimmer, M., Petersen, J., Sivyer, D. B., Mills, C., Young, E. and Parker, E. R. (2005) 'Impact of long-term benthic trawl disturbance on sediment sorting and biogeochemistry in the southern North Sea', *Marine Ecology Progress Series*, vol 298, pp79–94

Turner, R. K., Paavola, J., Cooper, P., Farber, S., Jessamy, V. and Georgiou, S. (2002) *Valuing Nature: Lessons Learned And Future Research Directions*, CSERGE Working Paper EDM 02-05. Available at www.uea.ac.uk/env/cserge/pub/wp/edm/edm_2002_05.pdf, accessed 29 March 2010

UK BAP (Biodiversity Action Plan) (1995) *Biodiversity: The UK Steering Group Report – Volume II: Action Plans*, December 1995, Tranche 1, vol 2

UK BAP (1999) *UK Biodiversity Group Tranche 2 Action Plans – Volume V: Maritime Species and Habitats*, October 1999, Tranche 2, vol 5

UK CEED (Centre for Economic & Environmental Development) (2000) *A Review of the Effects of Recreational Interactions within UK European Marine Sites*, Countryside Council for Wales, UK Marine SACs Project

Valette-Silver, N. J. (1993) 'The use of sediment cores to reconstruct historical trends in contamination of estuarine and coastal sediments', *Estuaries*, vol 16, no 3B, pp577–588

Van Dover, C. L., Grassle, J. F., Fry, B., Garritt, R. H. and Starczak, V. R. (1992) 'Stable isotope evidence for entry of sewage-derived organic material into a deep-sea food web', *Nature*, vol 360, no 6400, pp153–156

Virginia, R. and Wall, D. (2001) 'Ecosystem function, principles of', in S. Levin (ed) *Encyclopaedia of Biodiversity*, Academic Press, San Diego, pp345–354

Vorberg, R. (2000) 'Effects of shrimp fisheries on reefs of *Sabellaria spinulosa (Polychaeta)*', *ICES Journal of Marine Science*, vol 57, pp1416–1420

Worm, B., Barbier, E. B., Beaumont, N., Duffy, J. E., Folke, C., Halpern, B. S., Jackson, J. B. C., Lotze, H. K., Micheli, F., Palumbi, S. R., Sala, E., Selkoe, K. A., Stachowicz, J. J. and Watson, R. (2006) 'Impacts of biodiversity loss on ocean ecosystem services', *Science*, vol 314, no 3, pp787–790

Chapter 5

Review of Existing International Approaches to Fisheries Management: The Role of Science in Underpinning the Ecosystem Approach and Marine Spatial Planning

Andrew J. Plater and Jake C. Rice with contributions from Gillian Glegg, Sture Hansson, Manos Koutrakis, Stephen Mangi, Ivona Marasovic, Charlotte Marshall, Tim Norman, Temel Oguz, Frances Peckett, Sian Rees, Lesley Rickards, Lynda Rodwell, David Tudor and Nedo Vrgoč

This chapter aims to:

- Provide an overview of the data and modelling tools with which the scientific evidence base is able to support assessment, decision-making and adaptive marine planning and management;
- Discuss the challenges to implementing an ecosystem approach (EA) utilizing the available evidence base, tools for assessment and the current framework for developing shared knowledge; and
- Consider how the ecosystem approach may be facilitated by existing methodologies applied within a marine spatial planning (MSP) framework.

An ecosystem approach to management

Impacts on the marine environment

The marine environment is vulnerable to a spectrum of direct and indirect impacts arising from human activities both at sea and on land (see Table 5.1). As 'downstream' recipients of degrading impacts caused by poor land-use and catchment management, and simultaneously under increasing pressure to provide resources and space to meet human needs, the world's coasts and shallow seas are affected both directly and indirectly (Agardy, 2000) (see Box 5.1). It is, however, recognized that human impacts, both near- and far-field, need to be reduced to a level where marine ecosystems can function without serious impairment. This is particularly evident in fishing where past practice has over-exploited many marine fisheries and where in some marine ecosystems most stocks supporting traditional fisheries are in a state of collapse. The Food and Agriculture Organization's (FAO) *The State of World Fisheries and Agriculture* reports, for example, conclude that nearly 20 per cent of fish stocks globally are collapsed, over-exploited or recovering from past over-exploitation, and another 60 per cent are fully exploited (FAO, 2009). Similarly, where regional extirpations or near-extirpations have taken place, the cause is overfishing and habitat degradation together with biological and ecological factors (Powles et al, 2000).

In addition to target fish stocks, fisheries exploitation affects communities of organisms, ecological processes and sometimes entire ecosystems. The resulting impacts on biodiversity may be difficult to reverse, especially if fisheries impacts include habitat damage, and may take several years if not decades. If not managed effectively, commercial fishing, whether large or small scale, tends to over-exploit stocks, in some cases causing trophic mining (for example, Pauly et al, 1998) whereby decreases in the abundance of valuable species high in the food chain lead to the targeting of less valuable resources at lower trophic levels. As

Table 5.1 Threats to marine ecosystems

Type of Threat
Habitat loss or conversion
Coastal development (ports, urbanization, industrial sites, tourism)   Destructive fishing practices   Coastal deforestation   Mining (coral, aggregates, minerals, maintenance dredging for navigation)   Civil engineering works   Environmental change due to civil strife   Aquaculture-related habitat conversion
Habitat degradation
Eutrophication caused from land-based sources, for example, agriculture, sewage, fertilizers   Pollution: toxics and pathogens from land-based sources   Pollution: dumping and dredge spoil, also shipping-related   Salinization of estuaries due to decreased freshwater inflow/sea-level rise   Alien species invasions   Global warming and sea level rise
Over-exploitation
Directed take (low value, high volume) exceeding sustainable levels   Directed take for luxury markets (high value, low volume) exceeding sustainable levels   Incidental take or by-catch

Source: After Agardy (2003)

Box 5.1 *The Baltic Sea in a 100 years perspective*

The Baltic Sea has nine coastal countries, all but Russia being members of the EU. More than 85 million people in 14 nations live in the drainage area. The sea is used for very diverse and in some cases conflicting purposes: as a recipient for sewage, shipping, tourism, recreation, commercial and household fisheries, angling, and so on. Marine research and monitoring has a long history in this species-poor brackish water, providing substantial background for understanding and management.

The most prominent impacts of human activities have been caused by toxic substances, eutrophication and fisheries. In the future we can expect effects of climate change and acidification. Toxic substances (DDT and PCB) almost eliminated two top predators: gray seal and white tailed eagle. In the first half of the 20th century, hunting reduced the seal population by 80 per cent. In the 1960–1970s, toxic substances more than halved the remaining population. Reduced concentrations of toxins allowed the population to increase rapidly and it has grown to the same level as before these substances became a problem. The white tailed eagle also suffered from toxic substances, but the reproductive success has increased and the population size is now almost ten times as high as in the late 1970s.

A century ago, in a Baltic Sea that had lower primary productivity (in other words, not eutrophicated), seals consumed 300,000 tons of fish annually. Today's fishery catches 800,000–900,000 tons annually. It is, thus, realistic that seals once were an important competitor to humans for fish, and that the culling in the beginning of the 20th century allowed for increased catches. With the increasing seal population, culling will reappear on the management agenda. The eagles are not competitors in the same way. On the contrary, some eagles have specialized on hunting for cormorants and fishers often considered these birds a pest.

The nutrient load to the Baltic has increased over the past 100 years. Higher phytoplankton productivity has increased sedimentation. Below the permanent halocline, at around 70m, large areas suffer from hypoxia/anoxia, resulting in impoverished benthic fauna. Above the halocline the increased sedimentation of phytoplankton has allowed for increased biomasses of benthos. Sewage treatment plants have locally improved the water quality. Today, most of the nutrients come from diffuse sources – land run-off and atmospheric deposition. In general, there is no sign of reduced eutrophication and large quantities of nutrients in the sea will make the recovery to a more oligotrophic state a slow process taking many decades.

The reproduction of cod in the Baltic takes place in a few deepwater areas where the salinity is sufficiently high. Eutrophication has increased the frequency of hypoxia in these sites and over the past 30–40 years there is a clear correlation between deep water oxygen conditions and the reproduction success. However, in spite of good reproduction conditions in the beginning of the 20th century, the cod population was small compared with today. Ecosystem modelling suggests that this was because of lower productivity (less food) and predation by seals.

Fishing means killing of wild animals. Acknowledging this, it becomes obvious that fishing is a potentially environmentally devastating activity. Even if managed sustainably, fishing results in substantially decreased fish stocks. This, and the fact that fish constitutes a large fraction of the biomass at higher trophic levels, explains why fishing may have significant ecosystem impacts. The major target species in Baltic fisheries are cod, herring and sprat. For years, many of the catch quotas for these have been set much higher than recommended by scientists. Populations of cod and herring have thus been seriously overfished. Cod was intensively exploited already 40 years ago. In the late 1970s and early 1980s, catches doubled in response to increased abundance. After that, intensive fishing was maintained in spite of reproduction problems. This resulted in a historically low population in 2005. This mismanagement resulted in economic losses of about 100 million euros annually since the mid-1980s.

Sprat, the primary prey fish of cod, increased when cod decreased. The population was then reduced by an intensive fishery, but the stock is still at a high level. Herring decreased by 75 per cent during the last 25 years of the 20th century. Two factors appear to have contributed to the decline in herring. The first is catch quotas that seriously exceeded scientists' recommendations. Modelling results and decreased growth in herring suggests that food competition with the increased sprat population also contributed to the decline. Since the turn of the century, herring catch quotas have been in-line with the scientific advice, resulting in a stock recovery. In recent years, cod fishing intensity has been reduced and for 2009 the quota follows the scientific recommendation. This, and some years with improved reproduction, has resulted in an increased cod population.

Long term data on abundances of fish, zoo- and phytoplankton indicate that fisheries have impacted not only the fish stocks, but cascaded through much of the ecosystem (Casini et al, 2008). When the fishery reduced, the cod stock and sprat increased – this resulted in an intensified predation on zooplankton, and these decreased in abundance. Decreased zooplankton densities reduced the grazing pressure on phytoplankton, and the biomass of phytoplankton appears to have increased. This means that the mismanagement of the cod fishery may have changed the ecosystem in ways that amplified the effects eutrophication.

The actions against toxic substances have been successful. Currently the challenge is to prevent new substances from causing problems. Sewage treatment plants have reduced nutrient loads locally, but generally there are no signs of reduced eutrophication. Fish production has increased in response to eutrophication and a relevant question is whether we really would like to see a much more oligotrophic Baltic. Important fish stocks have been overfished, but in recent years fishing intensities have dropped and herring and cod have increased. If future quotas are set according to scientific advice, it is realistic that within one decade we will have a Baltic Sea with large fish stocks and good catches of fish with low concentrations of toxic substances. This may even reduce some of the effects of eutrophication. However, the potential impacts of climate change and acidification are open to speculation.

Sture Hansson, *Stockholm University*

requirements to provide for global food security increase, there is a need to harvest more, so more trophic levels are exploited by fishing through the food web. A decline in abundance of primary consumers then removes important forage species for organisms higher in the food web (Jennings and Kaiser, 1998).

Fishing methods commonly used to catch highly valued species selectively affect many other species (Dayton et al, 1995), for example, surface long-lining may contribute to the death of seabirds and turtles. Similarly, habitat alteration by bottom trawling may kill benthic plants and animals, and interrupt key ecological processes (Auster, 1998). Frid et al (2006) summarize the influences that fishing activities have on marine ecosystems, as:

- Direct removal of target species;
- Direct changes in the size structure of target populations;
- Alteration in non-target populations and communities of fish and benthos;
- Alterations in the physical environment;
- Alterations in the chemical environment;
- Food-chain effects, such as trophic cascades, and altered predation pressure.

From a UK perspective, the most important human pressure on the marine environment in terms of spatial extent and level of impact is fishing (Collie et al, 1997; Rijnsdorp et al, 1998; Dinmore et al, 2003; Eastwood et al, 2007). As a consequence, the development of a marine spatial planning (MSP) framework in UK waters requires detailed understanding of fishing pressure at the spatial scale of both regions and marine landscapes (Stelzenmüller et al, 2008). While this chapter draws from a wealth of research and experience in marine fisheries management, it is recognized that the issues considered and the role of science, monitoring and modelling in evaluating impacts on ecosystems and in underpinning multisectoral management strategies are applicable across a wide range of issues linked to human activity and climate change in the marine environment.

An ecosystem approach: Perspectives and policy

The increase in human pressures on the marine environment results in an increase in complexity of spatial use, requiring the protection of threatened and declining habitats (Douvere and Ehler, 2007). In the past, marine management approaches have been sectoral rather than resolving multiple-use conflicts (Stelzenmüller et al, 2008). Therefore, in recent years, emphasis has been placed on an ecosystem approach to integrated natural resource and environmental management (Douvere and Ehler, 2007).

There is distinction in some (but not all) jurisdictions between ecosystem-*based* management and an ecosystem *approach* (EA) to management. This is particularly the case in international settings where discourse takes place in languages other than English, and the two terms translate into more different concepts than they may seem. Ecosystem-based management is more prescriptive than EA. Ecosystem-based implies that the needs of the ecosystem come first, and management starts with as comprehensive a view of the ecosystem requirements as possible before addressing human needs. EA is more incremental and iterative, with past sectoral policies and practices being altered to accommodate ecosystem drivers and impacts if and where they may make a major difference. When a choice is made between one or the other, one is de facto choosing a position between these two camps. For example, FAO, supporting regulation of a major industry aiming to provide food security, generally takes an ecosystem approach, whereas the United Nations Environment Programme (UNEP), with a conservation mandate, generally talks about management being ecosystem-based. Here, we focus on the role of ecosystem goods and services in supporting an EA to management, and in particular the tools, knowledge and data that are used to underpin management decision-making in this context. However, this does not exclude the consideration of methods employed in ecosystem-based management practice.

It has long been recognized that fisheries management should be informed by an EA that places the species being managed in the broader context of the ecosystem (environmental, ecological and socio-economic) (OSPAR, 1992; Murawski, 2007; Shin and Shannon, 2010). This line of thinking has become evident in several international conventions and agreements, notably the 1992 Convention on Biological Diversity (CBD), the 1995 Jakarta Mandate on Marine and Coastal Biological Diversity, the 1995 Kyoto Declaration on Sustainable Contribution of Fisheries to Food Security and the 2002 World Summit on Sustainable Development. Managing the environment in an ecologically sustainable manner became a legal requirement through adoption of the CBD (Frid et al, 2005). In 1995, the FAO Code of Conduct for Responsible Fisheries (Garcia, 2000) provided a reference framework for incorporating ecosystem considerations into sustainable fisheries management (Shin et al, 2010a). Subsequently, the 2001 Reykjavik Declaration (FAO, 2002), the 2002 UN Sustainable Fisheries Resolution, the 2005 St John's Declaration and UN General Assembly resolution A/RES/61/105 on Sustainable Fisheries committed nations to implementing an EA to fisheries management. FAO states that fisheries should be planned, developed, and managed 'in a manner that addresses the multiple needs and desires of societies, without jeopardizing the options of future generations to benefit from the full range of goods and services provided by marine ecosystems' (FAO, 2003).

The critical role of healthy, productive ecosystems has been underscored more generally by the Millennium Ecosystem Assessment (MEA), stressing the linkages between ecosystems and human health and well-being (MEA, 2005) through the provision of both tangible and intangible goods and services, and in the marine environment by the 2005 UN General Assembly commitment to a Regular Process for Global Reporting and Assessment of the State of the Marine

Environment including Socio-economic Aspects (A/RES/60/30), and the work of the Assessment of Assessments to develop options for the Regular Process (UNEP and IOC/UNESCO, 2009). The current paradigm is, therefore, to maintain and/or enhance ecosystem function, status and health as the foundation of economic sustainability and growth.

EA promotes conservation and equitable, sustainable management of land, water and living resources (Frid et al, 2006). It relies on scientific understanding of ecosystem structure, processes, functions and interactions (CBD, 2000). Indeed, Cury and Christensen (2005) recognize that an EA to fisheries management requires integration of the spatial dynamics of the various components; and can be informed by quantification of the interactions between different components of the ecosystem. EA uses integrated planning tools such as strategic assessment, coastal zone management and MSP for regulating, managing and protecting the marine environment (Tyldesley, 2006; Boyes et al, 2007; Douvere et al, 2007). Crucial to successful marine spatial planning is the accurate assessment of spatial distribution of human activities and their associated pressures (Defra, 2005). There has, therefore, been a move internationally to bring sectoral management of fisheries into more integrated management of the marine environment (for example, Canada's 1997 Oceans Act; Australia's 1998 Ocean Policy; the US's proposed Ocean Policy; and the UK's Marine Bill) (Pascoe, 2006). The rationale for this shift is the need to take account both of the interdependencies among multiple activities that compete for ocean space, and of the impact one sector may be having on another (Barange, 2005). Some states and intergovernmental organizations (IGOs) support achieving this multisectoral integration through some form of super-agency with authority to integrate policy and management across sectors, but some other states and IGOs oppose establishing yet another hierarchical level of decision-making. Rather, they argue that sectoral management needs to be strengthened because it has a more complex job to do, and integration should be achieved by a framework that facilitates integrated planning but has no biding decision authority over participants (Rice and Ridgeway, 2009).

Europe has lagged behind other regions (such as North America and Australia) in developing a framework for integrated management (Symes, 2007) (see Box 5.2). The EU Maritime Strategy has underpinned the development of an integrated marine management which maintains ecosystem health while ensuring appropriate use of the marine environment for current and future generations (Rice et al, 2005). Recent consultation on the Marine Strategy Framework Directive (Directive 2008/56/EC) aims to establish this integrated ecosystem-based framework for environmental marine policy.

Integrated management of the marine environment requires the involvement of a broad range of stakeholders. These groups have different objectives, and may

Box 5.2 *Fisheries management in the Mediterranean: New approaches and perspectives*

The Mediterranean Sea has an area of 2.5 million square kilometres, which is 0.8 per cent of the total marine area of the world. The absence of upwelling of nutrients from deep waters in the euphotic zone and the relatively small amounts of discharge from land both result in low nutrient concentration in the euphotic layer. This oligotrophy is also reflected in the level of fisheries catches, in other words, 1.4 tons/km over the continental shelf. Moreover the trophic level of the Eastern Mediterranean Sea is lower than that of the western part (Stergiou et al, 1997). The Mediterranean has a narrow continental shelf with many biotopes, limited in extension and with different characteristics where a lot of marine species with overlapping biological cycles live.

Fish landings in the Mediterranean and Black Sea, grouped together by FAO, run to around 1.5 million tons/year, 30 per cent of which consists of small pelagic species fisheries (sardines/anchovy). These species live in large marine ecosystems and support large-scale fishing activities. Moreover they are the main prey of other important species such as tuna, hake, mackerel, and so on. The demersal fish species until 1995 had a significant part of landings from the EU countries, which from that year is on decline (FAO-Fishstat database).

Fisheries management in the Mediterranean

Mediterranean fisheries are characterized by fragmented fleets, usually composed by small vessels, use of large number of landing sites, multi-species catches and low catches per unit of effort. Most of the fish caught are recruits (0–1 years) of the main target species. Furthermore, no TAC or adaptive management is in place, so administrations do not require monitoring in order to manage fisheries. Until recently, no regular assessments were made by international working groups and the results of assessments made were rarely incorporated in management (Leonard and Maynou, 2003). Moreover, there are different fishing methods targeting the same species and considerable recreational fishing activity (more than 10 per cent of total landings).

European statistics indicate there are some 4300 larger 'industrial' or 'semi-industrial' units, mainly trawlers and purse seiners, operating from EU ports in the Mediterranean. However, the 'small scale' sector, including coastal and artisanal fisheries, is especially important. It is estimated that more than 40,000 small scale units are in operation. This is believed to be less than half of the total number of small vessels fishing in the Mediterranean. Described as a sea of small boats, fishing is part of the Mediterranean's cultural heritage, and for many communities remains an integral part of the way of life.

The main problem of the Mediterranean fisheries is overfishing, since the development of semi-industrial and industrial fishery has led to an overexploitation of many fishable resources. According to FAO, a number of Mediterranean stocks are overexploited, for example, bluefin tuna is the only species in the Mediterranean for which quotas have been set to address overexploitation. The impact of fishing gears to the marine ecosystems (for example, the impact on Posidonia beds) is another important issue (Chuenpagdee et al, 2003; Van Houtan and Pauly, 2007). Other problems include pollution, illegal and unreported fisheries, the low level of sectoral organization, impact from marine mammals (locally), non-scientifically documented legislation and technical measures for the Mediterranean (national and EU), low stock monitoring and limited scientific organizations and regional fisheries bodies for the monitoring of fish stocks.

Furthermore, for many exploited species, the rapid deepening of the Mediterranean leads to a spawning 'refugium' provided by a size/depth correlation whereby the large, old spawners live below the depth of many fishing operations. That historical refugium is now being placed at risk by new developments in gear technologies allowing fishing at greater depths.

European approaches and new perspectives

Following intense consultations with stakeholders the European Commission (EC) has proposed an ambitious reform of the Common Fisheries Policy (CFP), aiming to ensure sustainable fisheries in biological, environmental, social and economic terms. Albeit not specific to Mediterranean waters, the main challenges of this new CFP are: dwindling fish stocks; diminishing catches; too many vessels chasing too few fish; continuous job losses; and a lack of effective control and sanctions. To tackle fleet overcapacity, the EC basically proposes measures to reduce fishing effort. Furthermore, the EC wants to strengthen the application of the reformed CFP to the Mediterranean where circumstances differ greatly from those in northern fisheries. In the Mediterranean, catches are falling, the fish caught are getting smaller and some species are becoming rarer. Mediterranean fleets need to fish less and with less environmental impact, improve compliance with the rules, reinforce cooperation between fishermen and scientists and strengthen multilateral cooperation. It is important for the credibility and effectiveness of the CFP that the rules agreed by the council be applied correctly and in an equitable manner across the EU and on EU vessels in international waters. At present, the situation as regards declarations of Exclusive Economic Zones (EEZs) or Fisheries Protection Zones (FPZs) in the Mediterranean is inconsistent. The EC considers that the declaration of FPZs could be an important contribution to improving fisheries management, given that about 95 per cent of community catches are taken within 80km of the coast. These FPZs would certainly facilitate control and contribute significantly to fighting against illegal, unreported and unregulated (IUU) fishing. Moreover the EC, in order to conserve fishery resources, decided in 2000 to conduct scientific evaluations needed for the CFP (EU Regulations 1543/2000, 1639/2001, 1581/2004 and 199/2008). The data that were decided to be collected are on the fleets and their activities (number of vessels, gross tonnage, engine power, age of vessels, gear used, time spent at sea, fishing effort per category of vessel, per fishing technique and geographical area), on the main commercial fish stocks (through monitoring of catches, landings and discards and through surveys in the sea) and their biology and on economic and social issues.

Another valuable (also to fisheries) EU approach is Integrated Coastal Zone Management (ICZM), which aims to achieve sustainable use of coastal natural resources and to maintain their biodiversity, enhancing the social and economic prosperity of coastal communities and facilitating the interaction between different economic coastal sectors and the resolution of their conflicts (Clark, 1992). This can help to resolve conflicts of fisheries with a lot of other coastal zone activities, which even though they were developed recently have a strong economic and political power, for example, tourism, navigation, sport fishing and aquaculture.

In April 2008, the EC published a Communication on the role of fisheries management in implementing an EA to marine management. EA is the core of the EU's Marine Strategy, although other policy instruments for delivery of the ecosystem approach include the Habitats Directive (1992) and the Birds Directive (1979). In this text, the EC outlined how the CFP can form part of a more joined-up approach to protect the ecological balance of our oceans as a sustainable source of wealth and well-being for future generations. The basic objectives of the CFP include the application of the precautionary principle to fisheries management and the progressive implementation of an EA, but in practice a lot of work is still needed in order to implement an EA to Mediterranean fisheries management.

The exploitation of coastal zone fish resources through the development of artisanal fisheries appears mandatory for the Mediterranean (Farruggio, 1989). Some experts argue that small-scale fisheries, suitably governed – perhaps at community level – offer the potential for sustainable utilization of coastal resources (for example, Pauly, 2006), although others argue that it is not possible to 'suitably govern', in a coherent way, tens of thousands of fishers in thousands of boats fishing from hundreds of ports. However, although small-scale fisheries are potentially, and in many cases actually, more sustainable than large-scale fisheries, they may be disadvantaged because of remoteness, lack of infrastructure and marginal political power. Furthermore, small-scale fisheries may be disadvantaged when competing for fisheries resources and market access with heavily subsidized industrial fleets (Ponte et al, 2007). Moreover, artisanal fishery, even though it still represents the prevailing activity in many Mediterranean countries, has gone into a decline during the past half century that has relegated it to a marginal role, from an economic and social point of view, creating an erosion of customs and traditions (Charbonnier and Caddy, 1986). The recovery of artisanal fishery requires modernization by means of the reconstruction of both human and physical capital, currently subject to rapid ageing, but there is the danger that this modernization may eliminate those characteristics that make artisanal fisheries less harmful and more sustainable than industrial ones.

European waters have a very long history of local management. Collet (1999) points to continuum of a local management system in the Mediterranean based on the recognition of fishing territories and access regulation from the 3rd millennium BC, down through the centuries via the medieval guilds and 'brotherhoods' (Prud'homie in France, Confradias in Spain) to modern times. This tradition of local territorially-based management illustrates how the common pool resources of the seas can benefit from appropriate and effective management (Andaloro et al, 2002), although impacts at the time may have been kept low by limits on fishing technology or the mobility of fishers rather than the mode of governance. Many developing countries today use Territorial User Rights to communities with shares of the coastline, which provides incentives to manage *sedentary* stocks in a sustainable way.

The idea of imposing rights on coastal resources is also emerging as an approach for managing coastal fisheries. The problems of enforcement of regulations in the Mediterranean suggest that management-based incentives, or indirect incentives resulting from systems with increased fishers' involvement in the management process, would ultimately be more effective in controlling fishing. Output controls such as quotas, including individual quotas and individual transferable quotas (ITQs: transferred or traded TACs are set for each species in an area that is shared between boats) could be applied to certain fisheries since they offer proven advantages in allocating shared resources. Long-term quotas give fishermen a stake in conserving fish stocks, while open access is a free-for-all race to fish, which ultimately can lead to collapse. However, this approach could lead to serious equity problems for traditional fishing communities that have had customary rights of access and rights of harvest to local fish stocks.

In the Mediterranean, management is reactive, never adaptive and still less precautionary. There is a lack of feedback among the main agents for an adaptive management: administration, fishermen and scientists (Leonard and Maynou, 2003). To overcome the limitations, a set of measures should be implemented taking into account the international cooperation The General Fisheries Commission for the Mediterranean (GFCM), the International Commission for the Conservation of Atlantic Tunas (ICCAT), and other new Mediterranean bodies (such as EastMed, AdriaMed, SGMed, RCMMed) should play central role in such tasks. However, the more complicated management measures are, the more difficult enforcement will be and extensive surveillance will be required. Consequently, it seems less likely that small-scale artisanal fisheries with many vessels could deal with such measures. Moreover the participation of fishers in the crucial decision-making processes will be the decisive factor for policy enforcement and successful management. In this respect, the implementation of an ecosystem approach presents serious practical issues that need to be overcome, particularly in terms of engagement and enforcement.

Manos Koutrakis, *Fisheries Research Institute,*
Hellenic Agricultural Research Foundation

place different values on the components of the marine environment (Pascoe, 2006). They also have different risk tolerances for the costs of a decision on social, economic and ecological aspects of the decision (Rochet and Rice, 2005; Rice and Legacè, 2007). Further, while EA may provide a common overarching aim, there will be variation in management priority from one nation or region to another, linked to the relative importance of marine resources in sustaining a nation's or region's economy. This, therefore, presents potential for conflict when seeking agreement on multiple uses of specific seas where several sectors and/or countries make use of shared marine resources. This difference is particularly marked in the policy debate regarding marine genetic resources, where developed and less-developed States have strongly opposing positions (Ridgeway, 2009), but is also being played out in, for example, the south Pacific where states with high seas fisheries and states with an interest in ecotourism are often on opposite sides of debates about fisheries management (UN General Assembly 2009 Fisheries Resolution).

Economic valuation provides an important decision-making framework by potentially providing a common numeraire for the different stakeholders (Pascoe, 2006). For example, Balmford et al (2002) examine the value of retaining the undisturbed condition for a variety of ecosystems against that arising from conversion for economic development, as illustrated by the difference in benefit flows of goods and services. On this basis, integrated management is a process of resource allocation between competing users, employing the respective costs and benefits that can be optimized for different management priorities. Valuing the different uses of the marine environment (including non-marketed goods and services) is essential in order to ensure optimal allocation of the resources between competing sectors.

Marine resources, habitats and uses are located in various spaces and persistent over various times (Ehler and Douvere, 2009). Hence, successful integrated marine management needs practitioners who understand how to work with this spatial and temporal diversity. Most governments subscribe to some form of evidence-based marine environmental management and identify the role of scientific research as the sources of this evidence (Jennings, 2009). Science is used to understand how the different pressures exerted by humans affect states, and how the pressures would be altered by various management options, and therefore to predict the consequences of management action. The generally accepted view of the role of science in the provision of advice (Schwach et al, 2007) is that the science is an impartial source of objective advice to guide the overall political process of management (UNEP and IOC/UNESCO, 2009).

Climate change provides an additional challenge to understanding spatial and temporal patterns in marine ecosystems. Increasing CO_2 in the oceans and other physical impacts of climate change are likely to result in dramatic changes in marine and coastal ecosystems (Higgason and Brown, 2009). Ecosystem dynamics are controlled in part by external drivers, such as hydrodynamic/ atmospheric factors, which influence productivity of phytoplankton (Dickson et al, 1988; Richardson et al, 1998) and the dynamics of zooplankton communities (Krause and Trahms, 1983). Climate change and variability also affects the mortality, distribution and migration of fish populations (Jennings et al, 2001). Work thus needs to improve the reliability of predictions and to reduce the uncertainty in our existing knowledge of the effects of external drivers on fish stocks (Frid et al, 2006).

The precautionary approach has become a cornerstone of an improved approach to fisheries management (FAO, 1996; Richards and Maguire, 1998; Rochet and Rice, 2009): that action can be taken on current knowledge (from time series of data, laboratory analytical work, micro/mesocosm experiments and modelling) but this action must be supported by monitoring, thus tracking changes with which to evaluate progress in relation to agreed targets. In its simplest form, this implies that care must be taken to ensure that fishing is sustainable, and that uncertainty should be accounted for by reducing fish mortality (Stefansson, 2003). A precautionary approach to management requires the identification of ecological reference points against which management objectives might be set (Greenstreet and Rogers, 2006). According to Robinson and Frid (2003), there is a need to develop reference points for system-level emergent properties as measures of ecosystem health. However, one of the limitations of this approach is knowing which properties to use as metrics and in what way they will change under a given set of pressures and/or what the desirable value of the metric is (Hall, 1999; Rice 1999). Furthermore, achievement of objectives relies on the ability of scientists to evaluate and communicate the properties and functions of marine ecosystems, the ecosystem effects of fishing, and the effectiveness of management measures to maintain resources (Shin et al, 2010b).

The real challenge for management is determining the key limits – in other words, the types and levels of human activities that can be sustained without compromising the functioning of the ecosystem – thus providing the maximum sustainable take of target organisms with the minimum impact on other ecosystem components. This is very much a scientific issue (Frid et al, 2006).

Data and modelling in support of decision-making

Science contributes to the assessment of whether objectives are measureable, achievable or compatible and, if so, to the assessment of which combination of management actions can be used to achieve these objectives (Jennings, 2009). Science may also inform the state of the marine environment; the way in which ecosystem status is presented and interpreted may help society form a view about the choice of objectives and the trade offs needed to achieve suites of them. According to Frid et al (2006), however, two types of barriers inhibit implementation of ecosystem-based advice for fisheries management: (i) a deficiency in our scientific understanding of the ecosystem (including the human part of it); and (ii) methodological impediments to using the science in management. The precautionary EA, and to some extent MSP, requires fisheries and science to develop new or to adapt old tools to make them operational (Rochet and Rice, 2009). One area of new or adapted tools is the use of resampling methods that are designed to capture uncertainty more fully in assessment, so that science advisors and decision-makers know how uncertain the information base for a decision is (for example, Patterson et al, 2001; Berkson et al, 2002). The other group of tools uses simulation methods that allow the application of pre-agreed decision-rules (FAO, 1996; Garcia, 2005) and whose performance characteristics have been studied through simulation and use in fisheries management (Stokes et al, 1999; Punt et al, 2001; Butterworth and Punt, 2003). In this section, we consider the operational use of data and models in supporting EA to management of the marine environment.

Indicators

One of the challenges faced by the scientific community is to propose a generic set of ecological indicators that accurately reflect the effects of fisheries on marine ecosystems, and could support sound communication and management practices (Shin and Shannon, 2010). Indicators play two roles in management: (i) reporting on the effectiveness of past management actions to achieve the biological, social or economic objectives set (the audit function); and (ii) guiding decisions about the provisions of the management plan being developed (the control function) (Rice and Rivard, 2007). In fisheries management strategies, the emphasis is on using indicators in their control function: the control rule drives the science advice on the following year's management actions through the discrepancy between the current value of the indicator and some biologically based or societal reference point (Garcia and Staples, 2000; Defra, 2002; Rice and Rochet, 2005). For many years, fisheries management has focused on using a state indicator of spawning-stock biomass (SSB) and a pressure indicator of fishing mortality as the core products of assessments and the basis for management advice (Beverton and Holt, 1957; Ricker, 1975; Jennings, 2005; ICES, 2006).

In applying indicators, identifying reference points and setting objectives, an obvious requirement is that the indicators respond primarily to the anthropogenic activity being managed and are sufficiently sensitive that the impacts of the activity and the responses to management action are clearly demonstrable (Rice, 2000; Jennings, 2005; Greenstreet and Rogers, 2006; Rees et al, 2008; Blanchard et al, 2010). Indicators include size-based indicators (Shin et al, 2005) and trophic- (Cury et al, 2005b) or life-history-based indicators (Jennings et al, 1999; Greensteet and Rogers, 2006) and frameworks have been developed for their careful selection and evaluation of their relevance (Rice and Rochet, 2005; Rochet and Rice, 2005; Piet et al, 2008). Environmental and low trophic level indicators capture environmental change and bottom-up effects in a spatially explicit way, but the global influence of environmental change on higher trophic levels (for example, regime shifts) is not well represented (Cury and Christensen, 2005). Top predators or high tropic level indicators summarize changes in the fish communities, which are commonly related to exploitation. Top-down effects (such as trophic cascades) can be quantified using trophodynamic indicators; and several such indicators are required to measure the strength of the interaction between the different living components, and of structural ecosystem change resulting from exploitation (Cury and Christensen, 2005).

Keystone species might be regarded as important indicators in relation to ecosystem biodiversity (Cury et al, 2003). These are species 'whose impact on its community or ecosystem is large, and disproportionately large relative to its abundance' (Power et al, 1996). Keystone species often occur near the top of the food web because they have an impact on other species through consumption, competition and by modification of habitat characteristics. Ecosystems may also be described in terms of the feeding interactions among their component species (Pauly et al, 2002). The mean trophic level (TL) of fisheries landings (for example, algae at the bottom of the food web = TL1; large fishes whose food tends to be a mixture of low and high TL organisms = TL4) can be used as an index of sustainability in exploited marine ecosystems. Because fisheries tend to reduce the size of the fish in an exploited stock, they also reduce their TL. This is clearly illustrated in a study of the North Atlantic showing that the biomass of predatory fishes declined by two-thirds through the latter half of the 20th century (Christensen et al, 2001). TL does not change in direct proportionality to size composition of the fish community, so both size-based and TL indicators should be used together (Pope et al, 2006).

According to Murawski (2000), for EA to assume a greater role in management, unambiguous, quantifiable and predictive measures of ecosystem state and flux must be developed to index:

- Biomass and production by the ecosystem and relationships among its parts.
- Diversity at different levels of organization.
- Patterns of resource availability.
- Social and economic benefits.

One approach to integrating ecosystem-level information is the use of carefully selected and appropriate indicators to document ecosystem-scale impacts and allow the effectiveness of management measures to be assessed on these larger scales (Shin and Shannon, 2010). Aggregate indicators of ecosystem status include size spectra and biomass curves (Bianchi et al, 2000). Emergent property metrics may be derived from ecosystem models, or models of stability and resistance to perturbation of a food web (Rice, 2000). These are hard to use in support of decision-making because they integrate so much information that it is difficult to assess the effectiveness of alternative management options. While, in practice, a sound management strategy may be to employ a range of indicators, thereby providing 'weight of evidence' for evaluation condition and reducing uncertainty (Mayer and Ellersieck, 1986), the number of ecosystem indicators has expanded over the past decade (Cury and Christensen, 2005; Piet et al, 2008), threatening to confuse rather than to augment traditional single-species assessments and management approaches (Shin et al, 2010a). In this respect, the IndiSeas Working Group was established in 2005 under the auspices of the Eur-Oceans Network of Excellence to develop methods to provide indicators-based assessments of the status of exploited marine ecosystems. A comparative study of 19 fished marine ecosystems determined that a suite of ecological indicators is needed to resolve inconsistencies in information communicated by various indicators that measure different ecosystem attributes. The most constraining criterion in terms of selection was, unfortunately, that of data availability in the different ecosystems, rather than criteria related to the effectiveness of alternative indicators at either diagnosing ecosystem trends or guiding choices among management alternatives. Furthermore, indicators needed to be comparable across ecosystems, and estimation of the indicators not too costly (Shin et al, 2010a). The final set of indicators met these criteria, with at least one indicator per category (size-based, species-based, trophodynamic, pressure, biomass-related), and at least one indicator per management objective. Two indicators per management objective were selected (Table 5.2). The source of data for calculation of these indicators is diverse, including scientific surveys, records of commercial catches, stock assessment output, and estimates of species parameters (lifespan, TL).

Table 5.2 Ecological indicators selected by the IndiSeas Working Group, and the corresponding management objectives they meet

Indicator	Headline label	Used for State or Trend (S = State, T = Trend)	Management objective
Mean length	Fish size	S, T	EF
Trophic level of landings	TL	S, T	EF
Proportion of under- and moderately exploited stocks	% healthy stocks	S	CB
Proportion of predatory fish	% predators	S, T	CB
Mean lifespan	Lifespan	S, T	SR
1/CV of total biomass	Biomass stability	S	SR
Total biomass of surveyed species	Biomass	T	RP
1/(landings/biomass)	Inverse fishing pressure	T	RP

Management objectives: CB = conservation of biodiversity, SR = maintaining ecosystem stability and resistance to perturbation; EF = maintaining ecosystem structure and functioning; RP = maintaining resource potential.

Source: Shin et al (2010a)

Assessing the status of fish stocks can be difficult and prone to uncertainty; the task of assessing an ecosystem is far more challenging because there are no, or few, reference points at an ecosystem level (Jennings and Dulvy, 2005; Greenstreet and Rogers, 2006; Shin et al, 2010a), only incomplete data sets are available, and ecosystems are non-linear systems that can be difficult to model and to predict. Metrics of community structure have been proposed for evaluating the ecosystem effects of fisheries, although their application is not without reservation (Rice, 2000), and methods for estimating reference points on these metrics are largely untested (Rice, 2009).

Models

Single- and multi-species models
Strongly exploited fish populations will recover most, not all, of their previous abundance when released from fishing (Hardy, 1956). This is the basis for models of single-species fish populations whose size is affected by fishing pressure alone, expressed either as a fishing mortality rate or by a measure of fishing effort (Schaefer, 1954; Beverton and Holt, 1957). These models, still in use now in highly modified form, support the tuning of fishing effort to some optimum level to generate maximum sustainable yield (MSY). These evolved models can be run with only catch-at-age data, hence governments devote a large part of their resources to routine (and historically pragmatic) acquisition and interpretation of catch and age-composition data (Pauly et al, 2002). Traditional single-species assessment models suffer principally from their failure to estimate social and economic impacts, thus providing only part of the information needed for fisheries decision-making and inaccurate modelling of stock dynamics as a result of inaccurate or incomplete input data. Recent years have witnessed a significant shift in emphasis, with a growing recognition that single-species assessments are not sufficient for managing fisheries sustainably (Pikitch et al, 2004). Indeed, to make progress towards implementing EA to fisheries, integration of single-species assessments and holistic ecosystem assessments are needed (Shin and Shannon, 2010). Hollowed et al (2000) reviewed whether multi-species models are an improvement on single-species models for assessing fishing impacts. Although the benefits of multi-species frameworks include more realistic modelling of natural mortality and growth rates, caution must be taken in interpreting model output due to the inherent sensitivity to parameter estimation and the frequent assumption that populations exist in equilibrium. Furthermore, they concluded that most multi-species models address only a sub-set of the underlying factors that regulate ecosystem processes and that functional groups are invariably aggregated across different species or age groups. It is an

increasingly common practice that multi-species and partial ecosystem models are used to explore strategic questions about management policies and measures, but single-species models are still the basis for annual tactical measures such as quota adjustments. This allows management strategies to be evaluated in the context of any trends in the larger ecosystem context of the fishery, but annual harvests to use the most current information possible about the actual state of the stock(s) being harvested.

EA is predicated upon the need to assess broader fishery effects on the ecosystem, in other words, on the predators, competitors, the prey of exploited species, by-catch species and essential habitat. Therefore, the role of exploited species must be assessed within the ecosystem. The effects of changing environmental conditions on recruitment must also be understood and (if possible) predicted, so that the exploitation level can be matched to stock productivity (Koslow, 2009).

Ecosystem models

The International Council for the Exploration of the Sea (ICES, 2000) considered the insights that different models provide into how fishing may affect the ecosystem, according to seven resultant categories:

1. Habitat-based models – give an insight into how a population range will fit to the most suitable habitat as fishing changes the total habitat size.
2. Models based on community metrics – reflect how community-level metrics change in response to fishing disturbance.
3. Single-species models with variable prey or predators – incorporate one-way trophic feedback reactions on dynamic single-species models following a fishery disturbance.
4. Multi-species production models – show how fishery harvest or predator or prey will affect the abundance of the other.
5. Dynamic multi-species models – can incorporate spatially dynamic or age/size-structures populations into changes in predator–prey interactions brought about by fishing disturbance.
6. Aggregate ecosystem models – derive from food webs and energy budgets, illustrating changes in energy, carbon or biomass of aggregated functional groups.
7. Ecosystem models with age/size structure – distinguish from aggregate ecosystem models in that the individual functional groups are generally less aggregated and there is greater temporal resolution in their dynamics.

Box 5.3 *The Black Sea physical-ecosystem model*

The Black Sea in general, and its western coastal and shelf region in particular, have been severely degraded in the 1970s and 1980s due to rapidly intensifying eutrophication, chemical pollution, decline in living resources (mostly fish stocks), alien species invasions, and climatic variations. These have introduced marked changes in functioning and structure of the ecosystem. Consequently, ecosystem services to the community have been diminished considerably and their sustainability endangered.

Assessments of the present status of the Black Sea have shown that:

- The western coastal waters are still subject to high nitrogen enrichment.
- The pelagic ecosystem has achieved a reasonably good improvement with respect to the 1980s (but not to that of the 1960s).
- Climatic variations (cooling in the 1980s and warming in the 1990s) have played an important role on both biomass and species levels.
- The recovery of benthic ecosystem has not been encouraging so far, and is expected to take much longer time. Shallower regions (less than 30–40m depths) continue to be controlled by opportunistic species. But there are signs of improvement.
- Multi-species fishery is unsustainable. No recovery of predatory fish species. Some ongoing threats are: illegal fishing and destructive harvest techniques; lack of regional cooperative fishery management; and eutrophication-induced instability of the food web structure.

In the past, many previous attempts to manage individual threats in isolation have not been successful. This was mainly because marine ecosystems are complex adaptive and non-linearly interactive systems. This view has been supported by the recent qualitative and quantitative interpretations of the available data devoted to understanding individual and cumulative impacts of disturbances on human communities and marine ecosystems, especially over long (decadal) timescales. This implies that solutions to these problems demand a holistic approach both in terms of scientific and management perspectives. It further requires the implementation of an integrated system of social, legal, economic and ecological mechanisms and measures.

The fundamental basis for the Black Sea ecosystem model is the food web structure and how it is influenced by the external drivers of hydro-climatic control (physical, physiological factors), resource enrichment (eutrophication), over-exploitation and alien invasion, and internal interrelationships between different trophic levels of the food web, and a top-down control from predation. The physical-ecosystem model framework is driven by a physical model that simulates circulation and vertical structure in relation to seasonal stratification/overturning, and resultant changes in water quality (nutrients, ionic concentrations), which is coupled with ecosystem modelling of biomass, trophic interactions, larval dispersion, ecosystem dynamics, nutrient cycling, and so on.

The modelling plays an important role in ecosystem-based management of the Black Sea. Monitoring the full suite of ecological, economic and social indicators would certainly provide a comprehensive view of temporal and spatial changes in response to different natural and anthropogenic exogenous and endogenous disturbances, from changing political regimes to climate change. However, full scale, comprehensive monitoring strategy is not feasible under present economic capabilities of the Black Sea coastal and catchment countries. Hence, the development and implementation of end-to-end food web models is a necessary (but enormously challenging) task to accomplish. At present, highly non-linear dynamical marine ecosystems are difficult to model realistically especially at species and population levels due to lack of observed data to model physiological features. It is also important to consider trade-offs among social, legal, economic and ecological mechanisms and measures to maintain sustainable development of ecosystems, sustained delivery of their harvestable resources and provision of their goods and services to humans.

Other impediments to the development of ecosystem-based management of the Black Sea include lack of communication between natural and social science communities (which is critically important for building up a multi-scale, process-oriented perspective on the marine ecosystems dynamics), lack of political incentives from the coastal countries, and the lack of sufficiently trained personnel and uncertainties for long-term funding to sustain their activity.

Temel Oguz, *Institute of Marine Sciences,*
Middle East Technical University

Ecosystem models are now becoming established as management support tools. In terms of fishery impacts, dynamic ecosystem models provide an opportunity to make advances because they can both evaluate the state of the system and predict ecosystem response to different fishing scenarios (see Box 5.3). They also allow an examination of the behaviour of possible metrics such as a change in the energy flow or average trophic level (Robinson and Frid, 2003). In this respect, ecosystem models have been advocated as tools for the evaluation of system effects, but the extent to which models are able to make meaningful predictions has not been fully tested as yet. In addition, energy flow and average trophic level have a very high inertia and are not specific, so they give feedback only on the effectiveness of management measures on the scale of a decade at best. This alone is a serious shortcoming

Two commonly used models that incorporate fish stocks and interrelationships are Ecopath and Ecosim. The Ecopath with Ecosim (EwE) modelling approach combines ecosystem trophic mass balance analysis (Ecopath) with a dynamic modelling capability (Ecosim) for exploring past and future impacts of fishing and environmental disturbances as well as for exploring optimal fishing policies (Pauly et al, 2000; Christensen and Walters, 2004). Ecopath is the mass balance modelling component based on two master equations, one to describe production, and one for energy balance. The energy balance of a functional group is computed as consumption = production + respiration + unassimilated food. Ecosim is a time-dynamic simulation, consisting of biomass dynamics expressed through a series of coupled differential equations. The model predicts consumption, life-history dynamics, nutrient cycling and limitation, the movement and accumulation of tracers in foodwebs, fleet and effort dynamics, compensatory mechanisms, parameter sensitivity, fitting to time-series data, scenario-based testing of optimal fishing policy and close-loop simulations. Ecosim models can be expressed over a spatial grid (Ecospace) to allow exploration of policies while accounting for spatial dispersal/advection effects. In essence, Ecospace is the spatial simulation, dynamically allocating biomass across a user-defined grid map for predicting mixing rates, predicting spatial fishing patterns, advections, and seasonal migrations. However, as data are not always available to support user-defined parameters, there is some element of preconceiving the outcome of the spatial simulation and, hence, a confirmatory bias.

According to Christensen and Walters (2004), erroneous predictions from EwE usually arise from poor estimates for a few key parameters, rather than 'diffuse' effects of uncertainties in all the input information. Indeed, these ecotrophic multi-species models require many more data than single-species models. Also, owing to the complexity of the processes included, the uncertainty in the results obtained is difficult to quantify. In an objective critique of the EwE approach, Plagányi and Butterworth (2004) acknowledge some key strengths, but also shortcomings including the mass-balance assumption, the questionable handling of life history responses, over-compensatory stock-recruitment relationships, scale-related extrapolation problems, some mathematical inconsistencies in the equations, limitations arising from data quality and quantity, and inadequate consideration of uncertainty in input data and model structure. In addition, the lack of systematic testing of model behaviour cautions against the non-critical application of EwE. Nevertheless, EwE application has demonstrated the importance of certain processes in evaluating the impact of spatial management that may be missed in simpler models (Babcock et al, 2005).

Ecosystem models can be used to assess how changes in effort affect the ecosystem (Petihakis et al, 2007), but lack of data on fisheries components and linkages makes the implementation of ecosystem models such as Ecopath and Ecosim difficult in most areas. However, a variety of dynamic models have been developed over the past ten years, most focusing on trophic levels (for example, Hall, 1994; Tusseau et al, 1997; Chifflet et al, 2001; Crispi et al, 2001; Triantafyllou et al, 2003; Blackford et al, 2004). One of these models is the European Regional Seas Ecosystem Model (ERSEM) (Baretta et al, 1995). ERSEM dynamically simulates the biogeochemical seasonal cycling of carbon, nitrogen, phosphorus and silicon in the pelagic and benthic food webs, and is forced by irradiance, temperature and transport processes (Baretta et al, 1995). At the open boundaries time series are prescribed for dissolved and particulate nutrients, and the nutrient loads of rivers are prescribed at monthly intervals. A general circulation model is used to aggregate the exchange volumes across the box boundaries, from which horizontal transports of dissolved and suspended concentrations are calculated. The biological variables are represented as functional groups expressed in units of organic carbon, and chemical variables as the internal pool in the biological variables and as the dissolved inorganic pools in the water and sediment, expressed in units of nitrogen (N), phosphorus (P) and silicon (Si).

One thing that many models have in common is that the biological components are aggregated and abstracted into functional groups, for example, phytoplankton, zooplankton and nutrients. In ERSEM, the processes associated with each of the three types of organism are identified as: consumer, decomposer and primary producer (Blackford and Radford, 1995). The dynamics of the lower trophic levels are largely determined by the physical environment, which controls the resource supply, while the dynamics of the higher TLs may both be resource-controlled and predation-controlled.

With effective ecosystem simulation, management scenarios can be applied through key parameters (such as scenarios of reduction in fishing effort, parameterized through a reduction in benthic mortality parameters) and the impact can be seen in changes to other parameters (such as patterns of primary productivity). Although ecosystem models generally incorporate both benthic and pelagic variables covering a large percentage of the higher trophic groups, they often lack important extrinsic drivers, such a climate variation, which is fundamental for interpreting community patterns and dynamics (Hall, 1994). Other models deal in detail with only parts of the ecosystem. For example, the size-based model used by Duplisea et al (2002) focused on the impacts of trawling on benthic biomass and productions. Consequently, ecosystem simulation models may fall short of simple size-based models (for example, Pope et al, 2006) that describe the relationship between fishing mortality and the population size spectrum, and are more closely aligned to management questions.

Spatially explicit fishery simulation models

The focus on marine ecosystems has engendered the development of models that account for trophic interactions (for example, Walters et al, 1999; Watson et al, 2000; Pinnegar et al, 2005). However, the spatially explicit description of interactions between resources and fishing activities, including alternative management options, has received less attention (Pelletier and Mahévas, 2005). One reason is that most fisheries are complex systems, exploiting a diversity of resources (multi-species) with a variety of fishing activities (multifleet). Similarly, most fisheries models have a limited spatial description and do not account for the effects of environmental conditions on fish stock productivity or finer-scale distribution (even though changes in distribuition can affect fishery performance greatly). The generic and spatially explicit fishery simulation model ISIS-Fish (Mahévas and Pelletier, 2004; Pelletier and Mahévas, 2005) has been applied to a number of case studies in the North East Atlantic and Mediterranean (see Drouineau et al, 2006). ISIS-Fish was also used by Kraus et al (2009) to evaluate proposed and implemented fishery closures for the eastern Baltic. The Population model of ISIS-Fish comprises zones (such as spawning grounds, feeding areas), migrations (for example, movement between spawning and nursery grounds and feeding areas) and population parameters (such as age structure, growth curves). The Exploitation model includes data on catch and effort, fishing by various states, type of fishing gear, target species and catchability, while the Management model captures a range of management options.

Assessing the efficacy of a particular management tool – such as choice of mesh size or location of an area closure – cannot be done in isolation because the biological and economic effects of one tool are dependent upon how others are being used (Holland, 2003). To design an optimal mix of management measures, it is therefore useful to have a means of assessment that allows for concurrent evaluation of changes to at least the most important management measures. Holland (2003) describes the use of a spatial simulation model of Georges Bank cod that assessed nominal effort, mesh size and area closures – constructing revenue curves to explore the impacts on fishery productivity and profitability of different combinations under varying assumptions of spatial heterogeneity and the movement of juvenile and adult cod. The model was found to provide important insights into the relative impacts and utility of different management measures and helped identify and prioritize the critical data necessary to parameterize an operational management assessment model.

Management strategy evaluation (MSE)

Although not specifically supporting the delivery of an ecosystem approach to marine and fisheries management in an MSP context, management procedures (MPs) (Butterworth and Punt, 1999) and similar frameworks such as management strategy evaluation (MSE) (De la Mare, 1998; Butterworth and Punt, 1999; Smith et al, 1999; Sainsbury et al, 2000; Butterworth and Rademeyer, 2005; Schnute and Haigh, 2006) are becoming more widely used in fishery management because they provide formalizations of long-term, robust strategies that are designed to satisfy multiple conflicting objectives (Rademeyer et al, 2007), hence enabling a good level of integrated planning.

MPs assess the consequences of alternative management actions for both the target resource(s) and associated fisheries. The MP includes an assessment based on available data, as well as a method of translating assessment outcomes into management decisions. Simulation trials ensure that the associated decision rules lead to assessments that are robust to uncertainties regarding the dynamics of the resources being managed. The simulation framework essentially consists of an operating model (OM) to simulate the system of resource dynamics and fishery and generate future resource-monitoring data typical of what would become available in practice; an estimator that provides information on resource status and productivity from these data; and a Harvest Control Rule (HCR) that outputs the management action in the form of a Total Allowable Catch (TAC) or allowable fishing effort (Kell et al, 2006). The OM provides a means of conducting tests with a model designed to reflect realism and uncertainty in both population dynamics and the response by fishers to management implementation (Schnute et al, 2007). MSE simulations consist of first constructing an OM, assuming mathematical equations and parameter values for all processes from the ecosystem of the resource to the dynamics of the operating fleets. An observation model then represents the scientific observation of the system, from data collection to stock assessment, and an MP model combines this observation model with management decision-rules. Parameter uncertainty is addressed by replacing deterministic parameters with stochastic or probabilistic ones, and model uncertainty can be addressed by replacing a single functional relationship of interest with suites of alternative relationship formulations based on alternative assumptions and relationships.

The consequences of different management options are assessed by modelling several possible scenarios for the underlying true dynamics for the resource population(s) of interest and the impact of exploitation. Typical population dynamics models include age structure, growth, natural mortality and a stock-recruitment relationship with associated variability, but they may also include associated species or even the entire ecosystem (Smith et al, 2007). For models primarily of the populations exploited by the fisheries, the most uncertain parameters generally are about population productivity, as represented by the recruitment relationship with natural mortality (Rademeyer et al, 2007).

The MSE approach is a significant step forward compared with ignoring uncertainty and applying ad hoc decision-making (Rochet and Rice, 2009). MSE has all the benefits (and shortcomings) of any modelling exercise supplemented by tools to help make the precautionary approach operational. However, it is inherently impossible for MSE to provide the level of quantitative accuracy and precision needed to support finely differentiated management decisions. In addition, when a probability distribution is used to specify the expected value for a parameter, its use implies knowing more about a parameter than using a point estimate. Rochet and Rice (2009) emphasize that there are valuable uses of simulation-based MSEs, such as providing a process to define the management problem and consolidate knowledge, and eliminating 'bad' management strategies from consideration. MSEs should be used to strengthen the use of intelligence in developing approaches to fisheries management, not an excuse to remove it from the process. In this respect, the principle of Ecological Risk Analysis for Effects of Fishing (ERAEF) (Smith et al, 2007) offers an assessment framework spanning both population and ecological assessment, whose methods cover the full range from expert-judgement-based assessment through empirical model-based assessment.

Data time series

While data are costly to acquire and require high and often ongoing investment, there is no substitute to enable evaluation of current status, as well as recent trends or trajectories with which to monitor progress towards objectives (see Box 5.4) and facilitate adaptive management. Long-term data on unexploited populations are valuable for comparative (control) purposes, and the lack of such data has made it difficult to separate the effects of fishing from the effects of changes in environmental conditions (Stenseth and Rouyer, 2008). For example, in 2006, long-term data on larval fish abundance, the California Cooperative Oceanic Fisheries Investigations (CalCOFI), was analysed by Hsieh et al (2006), who concluded that fishing elevates variability in the abundance of exploited species. Further analysis of the 50-year CalCOFI data set shows that the mechanism causing the increased variation of harvested fish is the selective removal of the larger/older individuals leading to a decrease in average size and age of the fish (Anderson et al, 2008).

Data collection, collation, management and access are important issues in supporting decision-making in the marine environment. Similarly, frameworks and networks for the standardization and sharing of data represent important international collaborations that enable an ecosystem approach to make use of high quality data in a common form (see Box 5.5). The significance of this quality control and data availability cannot be overemphasized, as issues with inadequate or unreliable data have plagued previous work in attempting to identify reference conditions against which management can be assessed. For example, one approach towards understanding the impacts of management on marine ecosystems has been to examine temporal patterns in the ecosystem indicators discussed above to determine whether changes are consistent with expectations for sustainable use or over-exploitation. Blanchard et al (2010) used six indicators in a comparative study of 19 exploited ecosystems, showing that for most indicators trends were not detectable across a five-year period for several reasons: some ecosystems were already severely impacted; data were missing in some cases for recent years; the variance of each indicator was high; and the statistical power for detecting trends was low for indicator series less than ten years old (Nicholson and Jennings, 2004). In addition, time series spanning the longer period (1980–2005) were not available for sufficient ecosystems to

Box 5.4 *Spatial and temporal patterns in fisheries resources and ecosystem changes in the Adriatic*

The Monitoring of Croatian Fishery Resources in the Adriatic Sea national programme includes the monitoring of demersal and pelagic fishery resources, carried out through monitoring of bottom trawl, small pelagic and coastal small-scale fisheries. It is funded by the Croatian Ministry of Agriculture, Forestry and Water Management.

Bottom trawl fishery is one of the most important fishing activities in the Croatian part of the Adriatic Sea, both by quantity of caught organisms and by market value of the catch. The importance of this fishing activity is derived from the fact that the largest proportion of the catch is exported, primarily to EU countries and especially to Italy.

Figure 5.1 Biomass index of demersal species in different parts of the Adriatic Sea according to data collected during MEDITS expeditions

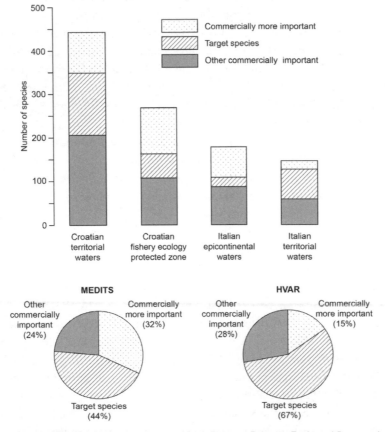

Note: HTM – Croatian territorial sea; EPH – Croatian Fisheries Ecological Protected zone; EPI – Italian extraterritorial sea, TTM – Italian territorial sea.

The main characteristics of demersal resources and demersal fisheries in the Adriatic Sea are as follows:

- Very large number of the species in the catches (multi-species fisheries). In the catches there are more than 100 different species with commercial value.
- Percentage of juveniles in the catches is very high (the majority of the catch is composed from individuals one or two years of age). Because of this, there are very high annual and seasonal fluctuations in biomass, mainly due to the differences in the intensity of recruitment.
- Demersal resources are exploited using numerous different kinds of fishing gears (more than 50 different fishing tools are in use), in other words, it is typical multigear exploitation. There are strong cumulative, competitive and synergistic effects of different gears.
- The most important fishing areas are located in the spawning and nursery areas of demersal species.

The majority of stocks are biologically common populations but economically shared between fishing fleets from different countries. Consequently, there is a strong need for harmonization of fisheries regulation and conservation measures between all countries participating in Adriatic Sea fisheries. Very big differences in the fishing fleet capacity in different parts of the Adriatic Sea have resulted in different fishing effort and, thus, the differential status of stock exploitation in the eastern and western part of the Adriatic Sea.

Figure 5.2 Distribution of chondrichthyes and commercially important demersal species in the Adriatic Sea during HVAR and MEDITS expeditions

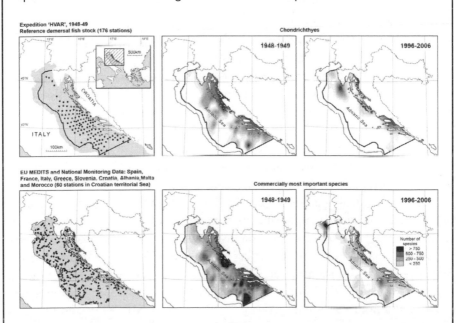

The recent state of demersal communities in the Adriatic Sea shows that they are subjected to high fishing effort that has caused pronounced adverse effects, including:

- Negative changes in the composition of demersal communities that are manifested in a reduction of the proportion of vulnerable species in the catch (long-lived species that have slow growth rates and low reproductive potential, for example, selachia and some commercially important osteichthyes).
- Reduction in total biomass of demersal species and reduction in biomass indices of a large number of the commercially most important demersal species.
- Changes in population structure of commercially important demersal species – primarily reduction in the mean size of caught specimens and size of first sexual maturity.

Generally speaking, the majority of commercially important stocks in the Adriatic Sea can be classified in the categories of 'fully exploited' and 'overexploited'. However, the state of demersal communities in different parts of the Adriatic varies and the degree of negative change is related to the level of fishing effort to which resources are exposed. The situation along the eastern Adriatic coast is that the Croatian territorial sea and inner sea is significantly better than along the western Adriatic coast and in the extra-territorial sea.

Ivona Marasović and Nedo Vrgoč,
Institute of Oceanography and Fisheries, Split, Croatia

carry out full cross-comparisons. Overall, the results were not encouraging in that there were no consistent patterns across ecosystems and indicators, making it difficult to identify broad-scale relationships between exploitation and ecological indicators or to generalize across systems.

These examples emphasize the shortcomings of past data collection and quality control and illustrate why monitoring of ecosystem attributes should be supported. Monitoring data provide the benchmarks against which model output can be tested and progress towards past and future management objectives can be measured. Further, the precautionary and adaptive aspects of both the ecosystem approach and MSP are predicated on the availability of evidence with which to assess progress towards management goals.

Data and modelling go hand in hand. Data may be interrogated to establish linkages and thresholds and enable understanding of ecosystem function. This understanding is the foundation of models, which may then be used predictively to evaluate the potential consequences of change in forcing, whether intrinsic (ecosystem dynamics) or extrinsic (due to environmental change and/or human action). Modelling then becomes an effective tool for scoping the potential outcomes of policy and management actions (such as MP, MSE), and helps to identify management goals against which success may be judged and, if necessary, actions redressed. Cury et al (2005a) previously recognized significant shortcomings in ecosystem data, including data availability and quality, uncertainty and spatial/temporal scales of operation. Acknowledging this, and indeed the caveats mentioned above in relation to ecosystem indicators, it is important also to recognize that data availability is improving and that it will continue to play a role in scenario testing.

Marine habitat mapping (MHM)

The ecosystem approach to management involves a set of interconnected concepts, disciplinary approaches, and technical issues that must be interpreted, synthesized and communicated across a wide community of scientists, policy-makers and stakeholders (Cogan et al, 2009). An important consideration for planning is the need to describe the spatial extent of seabed features and habitats (Stelzenmüller et al, 2008) through marine habitat mapping (MHM). Mapping of marine landscapes is important in determining the nature of biological communities and may assist in the designation of potential MPAs or areas of high biological value within a marine spatial planning framework (Roberts et al, 2003; Connor et al, 2006; Boyes et al, 2007). For example, mapping is useful for identifying areas that would be suitable for protection from the impacts of fishing gear (Fogarty and Murawski, 1998).

One of the first steps following the establishment of ecosystem-scale management goals should be to characterize the habitat features of the ecosystem. The process of MHM includes interpretation and classification of ship-based acoustic mapping of depth, substratum and geomorphology, in addition to (usually) satellite-based surface measures of roughness, currents, temperature and productivity (Cogan et al, 2009). Discrete spatial data themes that overlay habitat maps may then be used to represent some of the ecological processes operating at multiple spatial and temporal scales. This approach can be implemented in a variety of ways. The classification of the UK seabed into 44 types of marine landscapes was undertaken by the UK Joint Nature Conservation Committee (JNCC) (Connor et al, 2006), primarily using topographic and physiographic characteristics. In the US, the 2006 National Fish Habitat Action Plan (Association of Fish and Wildlife Agencies) involves two main components: (i) aquatic habitat mapping and classification; and (ii) habitat condition modelling whereby key components in the modelling include an index of biological productivity (closely related to biodiversity) and anthropogenic stressors. This pattern of habitat classification, mapping and modelling is increasingly used in support of marine management, such as in the US Gulf of Maine biodiversity assessment (Noji et al, 2008).

Box 5.5 *Data management in support of marine management in the UK*

Marine science frequently involves the collection of data. These data are often irreplaceable and are always unique, if only in the timing of collection. Even when considering all of the data collected, spatial and temporal coverage is sparse. Marine data can also be extremely expensive to collect. Over many years, a variety of databases have been compiled bringing together data from many different sources. More recently, there has been need for access to more multidisciplinary and integrated data sets to further our knowledge and understanding and to better manage the marine environment, including taking an EA and using marine spatial planning. In addition, there is an increasing requirement for operational data in near-real-time for forecasting marine conditions. Thus, it is of great importance to ensure that maximum benefit is derived from data once acquired. In the UK, we have adopted the slogan 'capture once – use many times' to promote this concept.

As far as possible, and for maximum use of data, the aim should be to make data freely available. To this end the World Meteorological Organization Resolution 40 and the Intergovernmental Oceanographic Commission (IOC) Data Exchange Policy (IOC Resolution XX-6) have been developed – both promoting free and open access to data. The International Council for Science (ICSU) notes that science has long been best served by a system of minimal constraints on the availability of data and information and promotes the principle of full and open access to scientific data. A strong public domain for scientific data and information promotes greater return from investment in research, stimulating innovation and enabling more informed decision-making. In line with this, the International Council for the Exploration of the Sea (ICES) has recently published a new data policy.

National and project data policies also exist and, as far as possible, they should be encouraged to provide free and open access to data. For example, the Irish Sea Coastal Observatory coordinated by the National Oceanography Centre, Liverpool, is a good example of a project which provides open access to its raw data in near-real-time and quality controlled data on a longer time scale. Against this, it is important to ensure intellectual property rights are not compromised and scientific papers produced by those responsible for the data collection. In addition it is important to give proper credit to the data collectors and data must be properly referenced or cited.

The British Oceanographic Data Centre (BODC) acts as the National Environment Research Council (NERC) Designated Data Centre for marine data and as the UK's National Oceanographic Data Centre within the IOC's international network of data centres. In addition to managing NERC's marine data, BODC acts as the data manager for the UK Marine Environment Monitoring and Assessment National (MERMAN) database and provides long-term stewardship and dissemination for data from the National Tide Gauge Network. However, even though BODC holds a large volume of diverse data, there is a wide variety of other marine data sets collected by UK organizations requiring long-term stewardship and improved access. For example, organizations such as the British Geological Survey (BGS) and the UK Hydrographic Office (UKHO) have large data collections of marine geology and bathymetry respectively stored in well-maintained archives. The Data Archive for Seabed Species and Habitats (DASSH) has been set up by Defra to cover a further data type. But many other data sets collected by government departments and agencies for particular purposes or projects have no long term archive assigned for the data.

The UK marine community is somewhat fragmented and it will take time to establish a fully productive partnership to improve the overall stewardship and access to UK marine data and information. However, significant steps have been taken towards this through the establishment of the Marine Environmental Data and Information Network (MEDIN), which builds on earlier similar initiatives. MEDIN is a collaborative partnership, open to all with an interest in development of a coordinated framework for management of marine data and information, and developing better access to the UK's marine data resources. Sponsors include government departments, research councils, environmental and conservation agencies, trading funds and commercial organizations.

During the pilot phase, a number of use cases were examined to assess metadata and data issues and requirements, including the Marine Spatial Planning Pilot Study and the Regional Ecosystem Study Group for the North Sea. These concluded that: sourcing existing data is generally difficult and time consuming; access to comprehensive metadata in a standard format remains highly variable; in general, marine data are not presented to a consistent format; and problems in accessing data and metadata are often repeated and it is often unclear how outputs can be made openly available due to licensing and payment problems associated with input data.

MEDIN aims to deliver a coordinated framework for the management of UK marine data including:

- Secure long-term management of priority marine data sets, according to best practice standards through a robust network of Data Archive Centres (starting with BGS, BODC, DASSH and UKHO).
- Underpinning UK and EU legislative and obligatory requirements for monitoring and marine planning (such as OSPAR, Water Framework Directive, European Marine Strategy Directive) in line with INSPIRE principles, and contributing to UK integrated assessments of the marine environment (such as Charting Progress) and the data needs of the newly established Marine Management Organisation.
- Improved access to authoritative marine data held in this network, through a central (discovery) metadata search capability thereby supporting its reuse and maximizing past investment in data.
- Improved evidence base for decision-making and marine planning built on best available data.
- An agreed set of common standards for metadata, data format and content, maintained and supported through implementation by partners.
- Guidelines, contractual clauses and software tools to support these standards and best practice data acquisition and management.

Lesley Rickards, BODC

Spatial data and geographic information systems (GIS)

The increased use of spatial measures in achieving ecosystem and community-level objectives (such as MPAs) requires the incorporation of spatial analysis into the calculation of population level indicators (Babcock et al, 2005). Geographic information systems (GIS) in coastal and ocean management are important because they can be applied to areas that are characterized by multiple spatial conflicts managed through a fragmented or sectoral framework (Valavanis, 2002). GIS have been used in fisheries applications for the past two decades as a monitoring, management and decision-support tool, particularly in conjunction with population dynamics modelling. Life history data include information of species type, preferred ranges of temperature, salinity and depth, spawning characteristics, migration habits, and so on. Spatial analysis is commonly used in nearshore environments for benthic habitat classification (Battista and Monaco, 2004). Stanbury and Starr (1999) developed a GIS for the Monterey Bay National Marine Sanctuary that allows manipulation of many terrestrial and marine data sets to create a database for the evaluation of natural resources. Incorporating data on hydrodynamics and fish, Valavanis et al (2002) developed a GIS of the cephalopod fishery in the eastern Mediterranean, which included habitat and oceanographic data as well as geo-referenced fishery data. Further, Lindholm et al (2001) used a dynamic model to investigate the link between survivorship of post-settlement juvenile cod and spatial variation in habitat complexity, simulating habitat changes based on fishing activities to evaluate the potential of an MPA for enhancing recruitment success. GIS have also been used in the study of many marine ecosystem information and analysis systems (Ault, 1996; Garibaldi and Caddy, 1998; Panzeri and Morris, 2000; Su, 2000; Varma, 2000; Megrey and Hinckley, 2001). Indeed, GIS technology contributes to fisheries management by processing fisheries monitoring data to spatially and temporally resolved information (for example, species geodistribution maps), integrating monitoring and environmental data, and providing integrated output to fisheries managers.

In designing marine management plans, it is important to take account of the spatial extent and patchiness of fishing activity, and the consistency with which areas are fished in the same region from year to year (Stelzenmüller et al, 2008). Descriptions of the spatial distribution of fishing pressures are more meaningful at a local level if they reflect habitat sensitivity to such pressures. Ecosystem-level indicators based on spatial analysis are only beginning to be developed. Fréon et al (2005), for example, have derived GIS-based indicators for the southern Benguela that include a spatial index of biodiversity, the exploited fraction of ecosystems and the mean distance of the catch from the coast. The availability of geo-referenced data on habitat, resource distribution and fishing effort, and the increasingly sophisticated methods available for visualizing and processing these data offer the potential to derive spatially explicit indicators of fishing impact from GIS (Babcock et al, 2005).

Visualization

Images help us to understand complex patterns, and also to describe, analyse, synthesize and convey information on a suite of indicators (Shin et al, 2010a). Visualization is thus a potent tool in the provision of decision support in fisheries information systems (Kemp and Meaden, 2002). Shin et al (2010b) describe a simple visualization tool to allow non-specialists to evaluate how heavily or lightly fisheries impact on an ecosystem. The approach emphasizes both identifying the ecosystem effects and communicating knowledge beyond the scientific audience, using ecosystem indicators to reflect key ecosystem processes and serve as signals that something more basic or complicated is happening than what is actually being measured (NRC, 2000). Kite and, in particular, pie diagrams were considered to be advantageous because they provide simple and multivariate summaries of ecosystems (Garcia and Staples, 2000; Pitcher and Preikshot, 2001; Haedrich et al, 2008). However, any form of visualization has to be interpreted carefully. Furthermore, fishermen, resource managers and scientists have different space–time perspectives, hence the trends and statistics they wish to derive from the information base must be specified as part of the visualization module interface.

Emerging technologies

Integration of data from acoustics and ocean-observation, as well as from satellites and other high-resolution oceanographic mapping tools, is likely to lead to major advances in fishery oceanography (Koslow, 2009). Acoustics can be used: to examine how topography and oceanographic features influence pelagic ecosystems; to characterize and map benthic habitats; to help understanding ecosystem dynamics; and for monitoring the effects of fishery closures. Multibeam side-scanning sonars have also been developed to observe near-surface schooling fish to determine structure and behaviour (Makris et al, 2006). Satellite remotely sensed oceanographic data provide reliable global ocean coverage of sea surface temperature, surface height, winds and ocean colour at high spatial and temporal resolution (Polovina and Howell, 2005). Altimetry data are used to construct regional indicators of the ocean vertical structure, such as ocean colour, sea surface temperature and to develop indices of biologically important ocean features.

Marine Protected Areas (MPAs)

This section aims not to review the principles, policy and practice of Marine Protected Areas (MPAs), but to illustrate how data, modelling and data handling considerations are brought together in their operationalization as a practical management tool for delivering the ecosystem approach (see Box 5.6).

Evidence shows that there are many circumstances where MPAs improve fish yields while conserving biological diversity more generally, providing they are correctly placed and implemented to reduce rather than redistribute fishing effort (Jennings and Kaiser, 1998; Agardy, 2000; Jennings, 2009). The important biological processes that support fisheries yield include spawning, migratory pathways, feeding settlement and concentrated feeding (De Groot, 1992); these processes are often concentrated in areas that can be identified by physical characteristics such as reef formations, extensive shallow areas, coastal wetlands, continental shelf breaks, and so on. Such identification is, therefore, a straightforward task for marine habitat mapping and GIS coupled with oceanographic data input. An additional role for MPAs is to serve as control sites for scientific research and experimentation. This adds value to adaptive management but is somewhat dependent upon a coordinated programme of data collection, management and communication, such as the ecosystem monitoring programme objectives of the Convention on the Conservation of Antarctic Marine Living Resources (CCAMLR) (Constable et al, 2000).

Used alone, MPAs mainly control the spatial distribution of pressures. Displacement of the activity causing a pressure is a key concern because the effects of a pressure outside the MPA could have greater or lesser consequences. A number of models have been developed to examine the effects of displaced fishing on target populations (for example, Horwood et al, 1998; Stefansson and Rosenberg, 2006), which can be assessed by combining information on habitat distribution, predicting changes in the spatial distribution of effort following the management action, and then modelling the impacts of fishing on habitat. Hiddink et al (2006) used this approach to look at the effects of alternative MPA designs on the aggregate biomass, production and species richness of benthic communities at the management region scale. MPA models tend to focus on single-species population dynamics (see review by Gerber et al, 2003). Generally, such models indicate that the effectiveness of MPAs in rebuilding or maintaining populations depends on the rates of immigration and emigrations, and on the fishing effort outside the reserve. Stock assessments have rarely included spatial analyses of effort distribution, although the need for it has long been recognized (Hilborn, 1985). Indeed, such analyses are important for designs that use both spatial and non-spatial management tools (Babcock et al, 2005).

MPAs, with 'no-take' reserves at their core, combined with effective effort limitations in the remaining fishable areas, have been shown to have positive effects in helping to rebuild depleted stocks (Murawski et al, 2000; Roberts et al, 2001). In most cases, successful MPAs have been used to protect rather sedentary species, rebuild their biomass and eventually sustain the fishery outside the reserves by exporting juveniles or adults (Roberts et al, 2001). Focused studies on the appropriate size and location of marine reserves and their combination into networks, given site-specific oceanographic conditions (for example, Sala et al, 2002), are therefore required (Pauly et al, 2002).

Box 5.6 *Marine planning tools: A case study of Lyme Bay*

On the 11 July 2008, the UK government closed 60 square nautical miles of Lyme Bay to scallop dredges and heavy demersal trawls. The aim of this closure was to protect marine ecosystems against damage caused by towed benthic gear. The area remains open to sea anglers, scuba divers, fishers using pots and nets, and other recreational uses. In light of the new Marine and Coastal Access Act 2009 and the future implementation of marine conservation zones in the UK, Lyme Bay is a very appropriate case study for researching the potential ecological, economic and social impacts of the closure of areas as marine conservation zones. Currently the Marine Institute (University of Plymouth) leads a Defra-funded project on Lyme Bay looking at social, economic and ecological impacts of the closure. Further ongoing Lyme Bay studies include environmental valuation of recreational use (Rees et al, 2010a, b), site selection for marine conservation zones (Peckett et al, 2010) and predictive modelling of species distribution (Marshall et al, 2010). In each of these studies, one or more appropriate marine planning tools is applied with the aim to better inform the marine planning process.

Marine planning tools

The emphasis on marine spatial planning in the Marine and Coastal Access Act highlights the need to establish appropriate tools for successful development of marine plans. In the case of Lyme Bay, the tools developed and employed include:

Cost Benefit Analysis (CBA) – facilitating the planning process by identifying potential cost and benefits and crucially the winners and losers of any policy decision. As well as looking at the scales of costs and benefits it is important to assess whether the costs outweigh the benefits and, if so, by what degree.

Environmental valuation – undertaken often as part of the cost benefit analysis to place a monetary value on environmental assets. This can be done using market or non-market valuation techniques.

Geographical information systems (GIS) – a tool which enables geographically dependent data to be displayed in an easily accessible and understandable visual format. It is a highly appropriate tool for simplifying the marine spatial planning process.

Marxan is a *spatial planning decision support tool* (DST) (Ball, 2000; Possingham et al, 2002; Ball et al, 2009) frequently used for the design of marine protected areas (Pattison et al, 2004). It uses numerical optimization to select appropriate sites for marine reserves if relevant habitat and species data are available.

Species distribution models (SDM; Elith and Graham, 2009) are simple in concept and identify relationships between the known distribution of species, or communities, and their present environment (Guisan and Zimmerman, 2000). The models can be applied to make informed predictions about suitable habitat in areas lacking distribution data. Applications of species distribution modelling in the marine environment over recent years include protection of highly mobile species (for example, Embling et al, 2010), optimization of catch per unit effort in pelagic fisheries (for example, Hazin and Erzini, 2008), and identification of areas to include in the development of future monitoring programmes targeting the recovery of benthic communities in a recently closed rocky reef area (Marshall et al, 2010).

Application of marine planning tools to the Lyme Bay case study

Many stakeholders are affected by the closure, for example, the fishing industry, fish merchants and processors, recreational users and adminsitrative agencies. In order to quantify costs and benefits of the Lyme Bay closure to the resource users a basic matrix of potential costs and benefits was first drawn up (Figure 5.3).

Impact of closure on fisheries (CBA, GIS)

In order to assess the socioeconomic impacts of the Lyme Bay closure on fisheries, Mangi et al (2009) applied both CBA and GIS. From the surveys, Mangi et al (2009) were able to determine the changes to income, total costs, time travelling to fishing sites, number of trips and fishing duration as perceived by the fishermen in the

Figure 5.3 Basic matrix of potential costs and benefits to resource users in Lyme Bay

Stakeholder	Cost	Benefit
Fishing industry	Displacement, change in gears, reduced landings	Scope for increase in static gear fishermen
Fish merchants	Sourcing fewer scallops, increased haulage costs	None
Recreational users	Potential for overuse	Scope for increased recreation, improved recreational experience
Administrative agencies	Increased enforcement costs	Contribution to UK MPA network

region. The main initial findings suggested either an unchanged income or a decrease; the decrease in income could be attributed to dredging less, increased fuel costs and displacement of towed gears into traditional fishing grounds of static gears.

Secondary data included fish landings data for ICES rectangles 30E6 and 30E7 (Source: Marine and Fisheries Agency), vessel monitoring system (VMS) and sightings data and enforcement costs from the deployment of surveillance aircraft and Royal Navy fisheries protection vessels by the MFA, and a patrol vessel by Devon Sea Fisheries Committee. Using secondary VMS data and GIS, the spatial distribution of different gear types in different years can be investigated. Different coloured dots can be used to show years and changes over time that may become evident (Figure 5.4). Furthermore, distribution of activity throughout the year can be analysed (Figure 5.5) or different fishing activities at a point in time.

Figure 5.4 Raw VMS data records for 2007 and 2008 by gear type

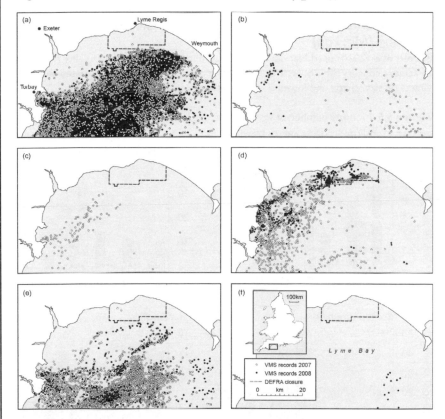

Note: a) Beam trawler, b) Pair trawler, c) Unspecified trawler, d) Demersal trawler, e) Scallop dredger, f) Bottom seiner

Source: Mangi et al (2009)

Estimating the value of recreation (valuation, GIS)

Questionnaires were developed for use online to determine the expenditure of divers and sea anglers in Lyme Bay and the turnover of dive businesses and charter boat operators as well as spatial distribution of their activities and perceptions of the closed area and their activity in the first year of closure. Respondents to the questionnaire were asked where they go and on a scale of 1–5 how frequently they visited the site. By summing up these data, 'hotspot' maps could be created showing relative frequency of visits (Rees et al, 2010a). The spatial data show how recreation activities vary across Lyme Bay. Alternatively, the spatial information can be used to try to estimate value of particular sites. Using the results from the survey we can estimate the value of sites inside and outside the closed area. The interesting thing is how these values may change over time, in other words, will protection increase the value of sites in closed area? Estimating a change in value can help in the evaluation of different policy options.

Assessing data requirements for marine conservation planning (Marxan)

Peckett et al (2010) used Marxan to help determine the level of data quality required to achieve conservation objectives in a marine planning process. They tested the effects on size, shape and location of sites selected of: using three different

Figure 5.5 Monthly number of sightings per day of observation for a) angling, b) potting, c) scallop dredging and d) trawling

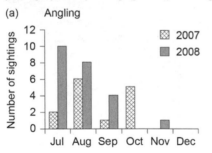

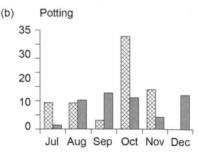

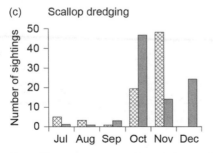

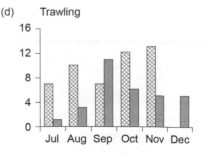

Note: Observations were only carried out in ICES rectangle 30E6 and not 30E7 in July–December 2007 and July–December 2008

Source: Mangi et al (2009)

resolutions of substrate layers; using data of increased complexity (biotope versus substrate) and; 'locking-in' the existing closed area to the solution (Table 5.3). The sites selected are then compared to those selected when no data are used in the selection process.

Using higher resolution substrate data resulted in increased protection for biotopes (Table 5.4 (a)). Using the no data layer, a data layer which allowed Marxan to choose randomly from the sites in Lyme Bay, resulted in more biotopes being protected by more than 20 per cent than the low resolution substrate. Using data of higher complexity, high resolution biotope data, led to at least 20 per cent of all biotopes in Lyme Bay being protected. As less complex data were used, the degree of protection was reduced (Table 5.4 (a)).

The locking-in of the closed area into the solution indicated an increase of up to 25 per cent in the solution in all of the outputs (Table 5.4 (b)) with the exception of the no data layer. This was expected because this output is totally flexible and only required any 20 per cent of Lyme Bay. The solution which appears to most effectively protect marine biodiversity in Lyme Bay is found when high resolution biotope data were used and there was no constraint to lock-in in the closed area (A1, Figure 5.6 and Table 5.3).

Table 5.3 Scenarios with Marxan using different data layers in the inputs, each scenario was run with the closed area locked-in and without it locked-in

Main data layer	Level of protection	Closed area not locked in	Closed area locked in
High resolution biotopes	protecting 20% of each biotope	A1	A2
High resolution substrate	protecting 20% of each substrate	B1	B2
Medium resolution substrate	protecting 20% of each substrate	C1	C2
Low resolution substrate	protecting 20% of each substrate	D1	D2
No data layer	protecting 20% of Lyme Bay	E1	E2

Source: Peckett et al (2010)

Table 5.4 (a) Number of biotopes (out of 22) protected by more than 20 per cent in the optimum solutions of each scenario, (b) Percentage area difference when the closed area is locked in to the output

	Number of biotopes protected in optimal solution	Percentage area difference if closed area locked in
High resolution biotopes	22	+25%
High resolution substrate	11	+15%
Medium resolution substrate	9	+20%
Low resolution substrate	4	+20%
No data layer	7	+4%

Source: Peckett et al (2010)

Figure 5.6 Lyme Bay best solution outputs for ten scenarios, with the existing closed area outlined in black. The filled-in areas indicate those planning units selected by Marxan for reserve

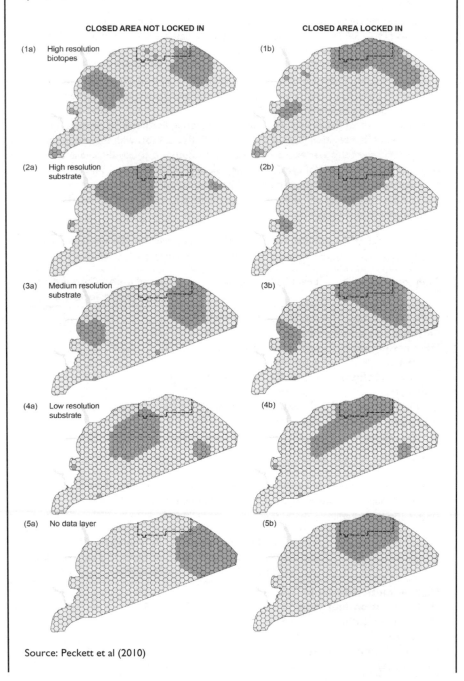

Source: Peckett et al (2010)

Monitoring recovery inside the closure: a species distribution modelling approach

In the Lyme Bay case study, Marshall et al (2010) set out to determine whether predictive modelling could be used to highlight locations within the closed area to prioritize for long-term monitoring of recovery, which is essential in order to assess performance against conservation objectives (Gerber et al, 2007). Prioritization of sites is important due to the expense, both financial and in terms of time, of monitoring and this is particularly important where long-term monitoring is required.

Binomial Generalized Additive Models and Generalized Linear Models were used to investigate the influence of environmental variables on the distribution of the pink sea fan *Eunicella verrucosa*; a fragile and long-lived member of rocky reef communities. Pink sea fan presence and absence data, obtained from towed video transects, were modelled against 13 environmental variables known to influence gorgonian distribution. Environmental variables with a significant influence on pink sea fan distribution were identified through the construction of bivariate models using each variable in turn. Following the stepwise selection of significant variables, the final model included only substrate although this alone explained 55 per cent of model deviance. The model performed well using a number of performance indicators, achieving a mean score of 0.76 for Correct Classification Rate, Sensitivity and Specificity (Fielding and Bell, 1997) following a threefold cross-validation.

Rocky areas were significantly and positively associated with sea fan distribution within the Bay (Marshall et al, 2010).

Lynda Rodwell, Sian Rees, Charlotte Marshall, Frances Peckett, Gillian Glegg and Stephen Mangi, *Marine Institute, University of Plymouth*

Hoffmann and Pérez-Ruzafa (2009) identify the general research challenges to MPA implementation as: strengthening the scientific basis for the selection and design; appropriately monitoring and evaluating effectiveness; studying the effects of MPAs in contrast to or in combination with other management tools; and formulating adaptive MPA design and management to cope with future climate change effects on habitats, species distributions and migration patterns. To this list of challenges should be added studies of the social and economic consequences of MPAs, and how to deal with the transitional costs. These themes of improved understanding and communication, effective assessment and adaptive management, and embedding a transdisciplinary approach in decision-making and implementation are further developed in the following section.

Challenges to implementation of the ecosystem approach to marine management

Given the comprehensive set of tools in the first section of this chapter, one might be forgiven for assuming that the evidence base for implementing an EA is comprehensive and that the tools and approaches can be readily translated into a support structure delivered through an effective MSP framework. However, significant deficiencies in our current science knowledge (Frid et al, 2006) include:

- Links between hydrographic regimes and fish stock dynamics.
- The importance of habitat distributions.
- 'Design rules' for MPAs.
- Ecological dependence/food web dynamics.
- Predictive capabilities in complex systems.
- How to incorporate uncertainty into management advice.
- The genetics of target and non-target organisms.
- The response of fishers to management measures.

Together, these limitations on our knowledge result in two major challenges to effective implementation of EA. The first is how the ecosystems, as a complex set of feedbacks and linkages, buffer and/or mitigate the initial ecosystem responses to the human pressures and management actions to mitigate them, thus adding a further complexity to the predictions (Frid et al, 2006). The second is that, because humans are also part of the fishery ecosystem, successful management requires prediction of the response of our own species to the management regime implemented (Pascoe, 2006), improving our ability to evaluate the comparative costs to user groups of different management options, understanding how to allocate 'rights' to ecosystem impacts other than catches of target species, how to allocate accountability for ecosystem 'costs' of fishing, and so on. Many of these gaps in knowledge, behavioural understanding and methodology lie at the boundaries of traditional discipline areas and are in need of transdisciplinary approaches.

EA has taken the progression from single species and, then, multispecies assessments of fisheries further by acknowledging that fisheries are part of the environment and cannot be managed in isolation (Cury et al, 2005a). With respect to understanding the ecological part of the ecosystem approach, Frid et al (2005) advocate applying environmental impact assessments, generally following six steps: (i) assessment of the current environment and resources; (ii) description of the consequences of fishing; (iii) assessment of the significant effects; (iv) selecting and evaluating mitigating strategies; (v) taking a decision on appropriate management action; and (vi) monitoring and reviewing. The difficulty for scientists in this approach is to acquire and assimilate data from the complexity of parameters and linkages that characterize the ecosystem. Primarily, difficulties lie in assessing the consequences of fishing activities and what constitutes significant effects (Petihakis et al, 2007). Among the barriers, challenges and opportunities associated with the development and use of indicators, for example, Rees et al (2008) note that:

- Most indicators address single pressures or do not directly reflect cause-effect mechanisms. Challenges include the uncertainty of reference conditions, natural variability of ecosystems, and defining the spatial and temporal scales within which indicators will function.
- Ecosystem-based management is more complex than fishery management and must deal with a wider array of indicators and communicate effectively with a wider range of stakeholders and decision-makers.
- Scientists need to better understand how their information may be linked to societal well-being as well as to change in natural environments.

Indeed, there is now a general understanding that assessments that integrate the social and economic dimensions of a decision with the ecological dimension are a minimum requirement for supporting implementation of the ecosystem approach (UNEP and IOC/UNESCO, 2009).

Cury and Christensen (2005) note that the interpretation of indicators requires scientific expertise because of the potential errors and biases in their analysis. A key lesson from the IndiSeas working group is that scientific expert knowledge is critical to correct interpretation of state and trend indicators and for distinguishing the effects of other potential ecosystem drivers such as environment or other human impacts (Shin et al, 2010a). The expert science advisors who help in such interpretation need to be policy independent; and not just of governments but of the spectrum of special interests from conservation advocacy groups to commercial enterprises. In addition, those closest to a particular analysis are not necessarily going have the most balanced view of the messages the analysis contains. Similarly, given the uncertainty about many processes operating in the marine environment, no single form of expertise can produce the right, or even adequate, answers (Schwach et al, 2007). Consequently, considerable iteration, feedback and knowledge sharing between scientific expertise and various stakeholders in the management process are necessary to operationalize indicators as the basis for assessing the ecosystem impacts of resource use (see Box 5.7).

The development and monitoring of both ecological and socio-economic indicators play an important role in supporting the implementation of EA by assessing ecosystem status, the impacts of human activities, the effectiveness of management measures and communication of fishing impacts to non-specialists (Cury and Christensen, 2005; Jennings, 2005; Rice and Rochet, 2005). Thematically, the United Nations Educational, Scientific and Cultural Organization (UNESCO, 2006) identified three classes of ecosystem indicators (slightly modified):

Box 5.7 *Marine planning in an EA context: Canada*

Canada's Oceans Act (1998, http://laws.justice.gc.ca/en/O-2.4/) provided the legal incentive to create a process for integrated management of human activities in the sea, applying a precautionary approach and an ecosystem approach to policy formation and management. A period followed of staffing-up a new sector of the Department of Fisheries and Oceans (DFO), and consultation widely within government (among 16 federal departments, and with provinces, territories and municipalities) and with diverse stakeholders. The result was the Ocean Action Plan (www.dfo-mpo.gc.ca/oceans/publications/oap-pao/page01-eng.asp), which laid out a two-pronged framework for planning and for such integrated management. One component, led by the Science Sector of DFO, was to consolidate and make usable the knowledge basis for integrated management in an ecosystem context. The other component, led by the new Oceans Sector, was to establish the governance process for integrating planning and decision making among the multiple levels of government. Five Large Ocean Management Areas (LOMAs), located in Canada's three oceans and the Gulf of St Lawrence (Figure 5.7), were chosen as the first units for action.

Figure 5.7 The location of the five LOMAs identified as the first integrated management areas under Ocean Action Plan I

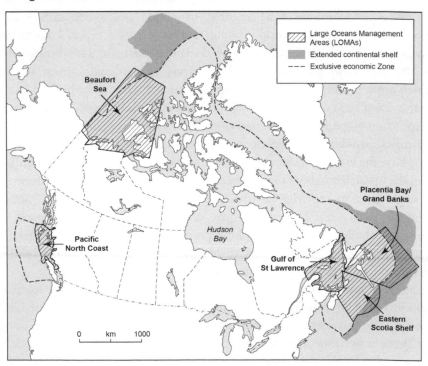

The initial plan was that the science component would undertake comprehensive ecosystem assessments of each LOMA. These would summarize existing knowledge of all ecosystem components from seafloor substrate and bathymetry to top predators (and all their interdependencies, as estimated by LOMA-scale models), coastal human demographics, the monetary and non-monetary values of social and economic activities in the LOMAs, and the impacts of those activities on the marine ecosystems. These assessments were soon abandoned for several reasons. They would be:

- a huge drain on science resources;
- incomplete in ways that limited their value in application;
- obsolete soon after completion; and
- equally boring to prepare and to read.

Instead, the knowledge component undertook the assembly of electronic data bases on major ecosystem components and human activities. Efforts were made to geo-reference core data sets from surveys, research projects, fishery monitoring, and so on, using a common geospatial software package chosen at the LOMA level. The geo-referenced data sets were augmented by meetings with holders of traditional and experiential knowledge; subsistence hunters in the north were relied on heavily for ecological information in the Beaufort LOMA, but experienced fishers provided valuable information in all areas.

A major advance in ensuring a degree of consistency across LOMAs in their evaluations of ecosystem status and trajectories was the decision to take a criterion-based approach to identification of four classes of special ecosystem properties: ecologically and biologically significant areas (EBSAs), ecologically and biologically significant species (EBSSs – but including community properties as well as individual species), depleted species, and degraded areas. The criteria for EBSAs and EBSSs were identified the same way. For each set of criteria, experts from across the country (government, academic and independent) were invited to national workshops. Several months before the workshops, there was a brainstorming conference call where every ecosystem property that someone thought might be a possible criterion was listed. Teams were formed on the call to prepare short literature reviews of the evidence why the property might be biologically or ecologically significant, and what operational problems might be encountered if it were to be used as a criterion against which places or species (or communities) would be evaluated. At the two workshops, these background papers were peer reviewed for adequacy of coverage of the literature and balance of treatment of the information that was extracted from the literature. In both cases there was little difficulty in gaining full consensus on a list of workable criteria and guidance on their use in assessments (Table 5.5, DFO, 2004, 2006a).

For depleted species, it was noted that under Canada's Species at Risk Act (2002), the Committee on the Status of Endangered Wildlife in Canada (www.cosewic.gc.ca) already conducts assessments of species potentially at risk of extinction in Canada, using the criteria and guidelines of the International Union for Conservation of Nature (IUCN, www.iucnredlist.org/documents/RedListGuidelines.pdf). Several provinces have provincial committees that do similar assessments at that scale, and DFO assessments of exploited fish stocks evaluate populations relative to abundance reference levels (DFO, 2006b). It was concluded that the combined list of all species identified by any of those three processes as either Threatened or worse under the IUCN criteria or below the fisheries reference levels would be considered 'depleted species'. For degraded areas, almost immediately there were objections from some provinces and municipalities that within at least three miles of the coast a federal department had no right to label areas as 'degraded', and legal advisors in DFO expressed concern that regardless of jurisdictional issues, once an area was called 'degraded' legal requirements for immediate remedial action would ensue. 'Degraded areas' was dropped from the first phase of the process.

Within this framework, regional meetings of experts, including a variety of mixes of academics, stakeholders, non-government organizations and participants from various levels of government, attempted to identify the ecologically and biologically significant areas and species in all five LOMAs. All five attempts were successful, although the mix of participants, quantity and quality of data sources, and time spent in preparing for these meetings was highly variable. The natural science and policy communities were most comfortable when the process had good monitoring data on ecosystem properties and economic activities. However, in cases when traditional or experiential knowledge from resource users was important to the overall information base for the system, formal engagement of social scientists skilled in extracting and validating experiential knowledge in culturally respectful ways were invaluable in the process. It was the combination of natural and social scientists that enabled a complete mapping of EBSAs and identification of EBSSs. Many of these meetings found that the majority of areas and species in the LOMA might be considered to qualify as at least 'relevant' to one criterion. Had the processes stopped there, the results would have been unhelpful to integrated (or non-integrated) planning and management, by producing guidance that every place and every species has some value and requires risk-averse protection. However, through a process of challenge and evidence-based dialogue, in every case the review and advisory meetings were able to proceed to a point where areas and species of *particular* significance to the ecosystem were identified; areas and species where management should exercise greater risk aversion than everywhere else, to avoid impacts that would be lasting in time or spread widely through the food web and/or in space.

Once the EBSAs, EBSSs and depleted species (from the Committee on the Status of Endangered Wildlife in Canada and fisheries assessments) were identified, the remaining science task was to integrate that information into coherent narratives of ecosystem status and trends, and integrate the ecological priorities with the social and economic information that was available on uses of the LOMA ecosystems. These documents, called Ecosystem Overview and Assessment Reports (EOARs), were to be the core science input to the next phase of the integrated planning process. It quickly became apparent that a set of additional guidelines were needed to ensure consistency in how the independently developed lists of EBSAs, EBSSs and deplete species should be merged. These guidelines were developed in another workshop/advisory meeting organized and conducted in the same way as the meetings to develop the EBSA and EBSS criteria. There were fewer precedents to draw from in this workshop, but full consensus on the guidelines available in DFO (2007) was reached with little difficulty. Aided by these guidelines, the EOARs were produced on schedule for every LOMA, identifying the priority 'Conservation Objectives' (matched to the EBSAs, EBSSs and depleted species) for each one.

Table 5.5 Criteria for ecologically and biologically significant areas, species and community properties

---

SIGNIFICANT AREAS

Main criteria

- UNIQUENESS – few alternatives to many alternatives
- AGGREGATION – exceptional density
- FITNESS CONSEQUENCES – site of activities which make major contribution to fitness of some species

Additional considerations

- RESILIENCE – highly sensitive/slow to recover
- NATURALNESS

Considerations in application

- Evaluations are RELATIVE not absolute
- Consider biological/ecological *FEATURES* in selecting EBSAs, and *THREATS* when choosing management measures for them
- Spatial and temporal scale, important and case-specific

SIGNIFICANT SPECIES AND COMMUNITY PROPERTIES

- Key Trophic Species
  - Forage Species
  - Highly Influential Predators
  - Nutrient importing and exporting species
- 3-Dimensional structure-providing species
- Properties above the species level
  - Size-based properties
  - Frequency distributions of diversity or biomass indices across species
- Additional considerations
  - Rarity and sensitivity

Source: After DFO (2004, 2006a)

In parallel with the science-led process to produce the EOARs and Conservation Objectives, the Oceans Management Sector was meeting with other federal departments, all levels of governments, all major industries, academics and citizens' groups from a broad spectrum of perspectives. The central theme of all these meetings was the same: what governance process would be the most effective for integrated planning and decision-making. From the outset, it was agreed that sectoral management at the various levels of government would remain. Sectors would continue to manage within their mandates; the integration was in planning and decision-making about the objectives to be achieved by the sectoral management. The consultations were to gain agreement on a planning process to set the integrated and mutually compatible objectives.

The vision was that the integrated planning 'tables' all legitimate parties (where 'legitimate' was defined broadly) would start with propositions of their preferred individual social (including cultural) and economic objectives. Once these were assembled from all parties, a dynamic interaction process with supporting science experts would use the information in the EOAR to provide at least general answers to a series of questions (Table 5.6)

Table 5.6 Questions addressed in the interaction process using information in the EOAR

Types of questions:	Concrete examples	
	Fisheries objective: a catch of 10,000 tons is needed to be economically viable	Recreational objective: 10,000 boaters want to use a specific stretch of coastline each summer
1. What parts of the ecosystem would have to be in what states in order for each objective to be likely to be achieved?	How large would the targeted stock have to be for a consistent catch of 10,000 tons to be sustainable?	Are there currently no boating obstructions, good navigational aides, and acceptable water quality?
2. Is the ecosystem currently in the necessary state, and if not, what would the transition costs be?	How low would the catch have to be for how long for the stock to rebuild to the necessary size in (1)?	How much would nutrient input have to be reduced for summer water quality to be acceptable to boaters?
3. What would the impacts be on the ecosystem if those objectives were being pursued?	What would the impact of taking 10,000 tons be on the present stock; and for the necessary effort what are expected bycatches and habitat impacts?	How much litter and oil leakage would 10,000 boaters produce; what types of coastal erosion from wakes, what types of facilities, and so on?

Once the first order answers to these questions were assembled for all the objectives of all legitimate parties, two further integrative questions would be posed:

4. Can all the objectives be achieved simultaneously? This identifies the inherent conflicts among diverse users. Examples include: Can recreational and commercial fishers both obtain the catches they 'need' from shared stocks? Can recreational boaters and coastal energy production facilities coexist in the shared space? Are the ecosystem impacts of a commercial fishery acceptable to conservation groups?

5. Can all the objectives be achieved without moving outside the domain of 'ecosystem space' defined by the Conservation Objectives from the science-based process that led to the EOAR? Examples include: Will the bycatch of a depleted species in a commercial fishery exceed the sustainable mortality rate for that species? Will 10,000 recreational boaters be likely to result in so much boat traffic that a cetacean population will be prevented from using a crucial feeding area at a crucial time for their life history?

The integrated management table was to be the place where the conflicts from questions (4) and (5) would be resolved. However, the conflicts captured by (4) and (5) are fundamentally different, and require discussions of quite different natures. Conflicts around question (4) are conflicts of those with different societal priorities and values. They require negotiation and compromise among different industries and stakeholders. There are no fixed boundaries; any of the parties can compromise as much as necessary to produce mutually compatible objectives. The winners and losers reflect society's preferences for some uses (or for the benefits from some uses) over others.

Conflicts around question (5) are conflicts between what different societal groups want and what the ecosystems can sustain. Society can always choose to act in ways that are unsustainable, but to the extent that the Conservation Objectives are reliably identified they are fixed boundaries. The ecosystem itself has no capacity to compromise without negative impacts on the services it can provide sustainably and there is rarely more than limited opportunity to move the Conservation Objective 'boundaries' through remediation or enhancement activities (which have their own costs to be borne). The negotiations are among the users of those ecosystems to see which combinations of uses can be simultaneously achieved while not *in aggregate* causing harm to the ecosystems that is serious and/or difficult to reverse (as per a number of international agreements, from the Rio Declaration – www.un.org/esa/dsd/agenda21 – onward).

Exactly how the complex process of these negotiations among users would occur, such that in the end options for compatible objectives for all user groups would result, was to be established once agreement on the nature of the 'integrated planning tables' was reached among the various levels of government and legitimate stakeholders. Such agreements have not yet been reached in any of the LOMAs. All parties recognize the value of these tables, but some impediments to progress have proven difficult to overcome. None of the departments and levels of government have been willing to accept any binding outcomes from these tables. Many industries have at least hinted of an inability or unwillingness to even state their objectives with the degree of specificity necessary to serve as the basis for negotiations even under question (4), let alone under question (5). Many stakeholder groups have hinted at the intent of using these tables to push for politicized and value-laden debates from the outset. This combination of different challenges from different parties to the tables has made it to this point impossible to gain agreement on who will sit at the tables under what conditions, and what 'rules of engagement' will apply. Combined with major funding challenges in recent budgets, progress on the governance system to use the information in the EOARs and Conservation Objectives is slow. The commitment to improve the integrated nature of planning and management for Canada's oceans remains high, but the path to take these final steps is not yet clear.

Jake Rice, *Fisheries and Oceans Canada*

References

DFO (Department of Fisheries and Oceans) (2004) *Identification of Ecologically and Biologically Significant Areas*, DFO Canadian Science Advisory Secretariat, Ecosystem Status Report 2004/006

DFO (2006a) *Identification of Ecologically Significant Species and Community Properties*, DFO Canadian Science Advisory Secretariat, Science Advisory Report 2006/041

DFO (2006b) *A Harvest Strategy Compliant with the Precautionary Approach*, DFO Canadian Science Advisory Secretariat, Science Advisory Report 2006/023

DFO (2007) *Guidance Document on Identifying Conservation Priorities and Phrasing Conservation Objectives for Large Ocean Management Areas*, DFO Canadian Science Advisory Secretariat, Science Advisory Report 2007/010

1. Ecological indicators: to characterize and monitor change in the state of various physical, chemical and biological aspects of the environment relative to defined quality targets with thresholds for management action (see also Fisher et al, 2001).
2. Socio-economic indicators: to measure whether environmental quality is sufficient to maintain human health, human uses of resources and favourable public perception (see also Cairns et al, 1993).
3. Governance indicators: to monitor the progress and effectiveness of management and enforcement practices towards meeting environmental policy targets.

In this respect, the metrics with which to assess fisheries or marine resources management extend well beyond the scope of much of the monitoring and modelling described earlier.

Communication and knowledge exchange is also a crucial aspect of marine resource management. Both common sense and international law underline the important role of science in management but in classical fisheries management, the key issue appears to be the control of removals at a level where stocks are capable of replenishing the losses (Hoydal, 2007). Here, the level of removal depends on the population dynamics of each stock and is set according to the best scientific advice, based on historical trends in productivity (such as growth and recruitment). Shelton (2007) cites weaknesses in the scientific basis for management decisions that led to the decline and collapse of many Canadian east coast groundfish stocks in the late 1980s and early 1990s (Hutchings and Myers, 1994; Walters and Maguire, 1996), which included too much emphasis on point estimates in earlier assessments and insufficient consideration of uncertainty and risk (Shepherd, 1991; Smith et al, 1993; Shelton and Rivard, 2003). In contrast, Rice (2005) recognizes a broader range of weaknesses in the linkage between the scientific evidence and management action in this case, for example, uncertainties in the evidence base (which increase during the onset of fish stock declines), asymmetrical management responses to the evidence (favouring good news), the response time of the science–management exchange process (with only swift action being able to stem declines), vulnerability of the process to politicization, and poor fishing practices being particularly devastating when the fish stock is at its most vulnerable. Although biological considerations are necessary preconditions for sustainable fisheries, economic, social and political constraints are equally important. The challenge is, therefore, to frame a scientific approach that offers real policy guidance, coming not only as a result of scientists' ability to analyse, predict and explain, but also by making the scientific arguments relevant, clear, credible, timely and accessible to the non-expert. Frid et al (2006) stress that science needs to be presented directly to decision-makers in a manner that allows them to engage in the dialogue. This engagement must occur throughout the management process. Decision-makers and stakeholders require sound knowledge so that the management objectives set are achievable and arrived at after due consideration of the options and alternatives. It then must continue as progress towards those objectives is monitored and evaluated, and choices about management responses (whether ad hoc or in formal adaptive frameworks) are made. Schwach et al (2007) examined the way scientific advice has been generated from technical and institutional perspectives, suggesting that many people involved in the fisheries management need new ways to engage science in management. For example, rather than playing the role of 'experts' who set next year's TAC, scientists can help facilitate interactions among stakeholders in trying to build an accurate picture of the marine environment. The themes of boundaries between science and policy, stakeholder engagement, and linking existing assessment processes are particular foci of the Assessment of Assessments report (UNEP and IOC/UNESCO, 2009). This emphasizes that, while a strong link between the assessment and relevant decision-making process is vital, there has to be a clear agreement that experts have the final word with respect to the factual analyses and interpretation. In contrast, experts may contribute to policy recommendations made by governments and stakeholders but they do not have a veto over these recommendations.

Spatial considerations also need to be more prominent in monitoring, modelling and decision frameworks. EA incorporates a range of objectives related to maintaining the structure and function of ecosystems and their components (Pikitch et al, 2004). Because some components have a restricted distribution, their protection may require spatial management (Babcock et al, 2005). In addition, spatial models of fleet dynamics allow more accurate predictions of the effects of envisaged management measures. Indicators such as diversity indices could also include a spatial component. Consequently, the role of GIS and spatial analysis becomes increasingly important in supporting management decisions.

Modelling should not be viewed as an alternative to expert judgement, discourse and deliberation. Few operating models, for example, can simulate the actions of scientists, fishers and managers making choices in each year of an ongoing cycle of fishery-assessment-advice-management plan-fishery (Rochet and Rice, 2009). Bayesian approaches (Punt and Hilborn, 1997) amplify many concerns about treating quantitative details as more meaningful than actually justified by the quality and quantity of data or theory (Rochet and Rice, 2009). Simulations (such as MSEs) have important roles in supporting management decision-making but their outputs are only one source of information to support decision-making rather than being a substitute for applying full intelligence and developing shared knowledge in the management process. In MPs and MSEs, robustness-test scenarios need to be identified (Cooke, 1999), reflecting the true dynamics that may vary more widely, be less plausible or have less impact than those considered in the reference set (Rademeyer et al, 2007). It is generally useful to compare both empirical and model-based MPs. The latter, when based on an age-aggregated population model, often prove a good choice, particularly over the longer term (Rademeyer et al, 2007). Furthermore, the performance statistics chosen to aid a selection of candidate MPs need to be meaningful to all stakeholders and careful thought needs to be given on how to best present these to permit easy comparison. Indeed, being able to visualize the implications of a scenario allows better formulation of future management strategies.

The marine spatial planning (MSP) toolkit for ecosystem-based marine management

This section illustrates how the process of MSP can incorporate current management practice, assessment methodologies, data and modelling frameworks, and mechanisms for transdiscplinary knowledge sharing from EA to fisheries management. MSP is adaptive and takes place by iteration (Ehler and Douvere, 2009). Consequently, monitoring, data gathering and modelling are essential support tools for delivering MSP, playing key roles in:

- Establishing the current situation (Step 5).
- Establishing the future situation (Step 6).
- Monitoring MSP (Step 9).

Marine resources, habitats and uses are located in various spaces at various times. Successful marine management therefore needs practitioners who understand how to work with these distributions in time and space (see Box 5.8). Successful MSP is often based around the tackling of specific problems or conflicting uses – existing or anticipated – and, hence, the goals are likely to be framed around current and/or emerging national priorities. The aim of MSP is to balance demands for development with the need to protect the marine environment (Ehler and Douvere, 2009); paralleling the fundamental basis for sustainable fisheries management.

Box 5.8 Moving towards integrated marine policy and planning: The Crown Estate Marine Resource System (MaRS)

The requirement for an integrated approach to the management of the marine environment is well-known and documented (for example, Cicin-Sain and Knecht, 1998; Foster et al, 2005). Integrated planning and subsequent management aims to achieve sustainable development of the marine environment and its resources through effective and collaborative processes, yet the complex nature of the marine environment, physically and legislatively, requires specific and appropriate planning and management that differs from the terrestrial environment.

The reasons for the lack of integration in marine policy, planning and management are numerous; they include:

- The complexity of responsibilities acts as a barrier to agencies and organizations taking an integrated approach (Shipman and Stojanovic, 2007).
- A lack of clear policy regarding the marine environment leads to poor integration among countries, at the coast and between regional and local scales.
- There is a lack of understanding that the oceans need specific clear policies and that their planning and management cannot simply be transferred from terrestrial policies or processes (although much can be learnt from terrestrial regimes).

In recent years, new legislation including The Marine and Coastal Access Act 2009, The Marine (Scotland) Act 2010 and the EU Marine Strategy Framework Directive has crystallized policy drivers for marine planning and its integration in the UK. This legislation and the emergence over the past decade of ever-increasing uses and pressures on the seas further emphasize the need for an operational integrated approach.

As owner of the seabed to the 12 nautical mile mark, and having rights out to the continental shelf, The Crown Estate is custodian of a marine environment more than 850,000km^2 in size; having responsibility for understanding marine and coastal environments and considering the best way to ensure long-term sustainable development, and being responsible for leasing many commercial activities in the UK waters. These include, among others, renewable energy (wind, wave and tidal power), extraction of marine aggregates; and, leasing space for cables and pipelines. As a land-owner not a regulator, the statutory process of planning the marine environment falls to relevant government departments (such as the Marine Management Organization or Marine Scotland). The integration of our own planning and decision-making with the government and its agencies is, therefore, vitally important and we take this responsibility seriously. In addition, as the majority owner of the intertidal area around the UK, we are also acutely aware of the need for integration and cooperation with local authorities and consideration of terrestrial planning frameworks, for example, the EU Water Framework Directive. The focus of our activity is to help us plan our commercial activities in a way that is also consistent with our stewardship objectives and, in due course, compatible with statutory marine plans when adopted.

MaRS

The Crown Estate is responding to calls for sharing of data and information and providing stakeholders with opportunities to influence decisions about the allocation of the seabed for different uses. Managing involvement of this kind is a challenge, particularly if the intention is to meaningfully engage stakeholders in decision-making rather than simply informing them of decisions that have already been made. To help their marine spatial planning team identify constraints and opportunities for future development, The Crown Estate has developed a GIS-based Decision Support System (DSS) referred to as the Marine Resource System (MaRS). MaRS integrates information from a database of more than 450 GIS layers, including data and information about:

- physical characteristics of the seabed, such as bathymetry and sediment type;
- environmental data, such as nature conservation designations;
- economic uses, such as value of fisheries, locations of existing lease areas, aggregates extraction areas; and
- natural resources, such as wind and current speed.

MaRS can be used simply to map existing conditions or the location of existing marine assets. It can also be used to identify the relative suitability of the seabed for different types of activity, including, for example, renewable energy developments, marine aggregates (sand and gravel) extraction or aquaculture. Users can characterize the suitability of the seabed by identifying the physical, technical and environmental factors that constrain or otherwise influence their construction or operation. In practice, this is achieved by users selecting those layers they consider to be relevant to the analysis, weighting them according to their perceived influence on the suitability of a location for the activity. For some activities, the presence of a feature is a clear constraint on development. For example, the presence of exposed bedrock on the seabed currently precludes the construction of offshore wind farms. For many other factors, the degree of constraint or opportunity depends on the extent to which a characteristic is expressed often within key thresholds. For example, offshore wind farms are typically constructed in relatively shallow water depths with relatively high average wind speeds.

MaRS uses multicriteria analysis (MCA) to combine the weighted layers and to generate maps indicating the relative suitability (or quality) of the seabed for the activity of interest. The outputs are relative and user defined, they reflect the interests of the user and their understanding of the factors influencing suitability. Different users are likely, therefore, to generate different outputs. Rather than being a disadvantage, this ability to generate user-specific output is an advantage, as it emphasizes the neutrality of the tool.

The Crown Estate is using MaRS to formulate an understanding of current and likely future uses of the Marine Estate. A key application has been the planning of future leasing rounds for renewable energy in response to government's objectives to expand these technologies in response to the challenges posed by climate change and the need for energy security.

Importantly, MaRS is also emerging as a place where marine policies can be explored and developed, for example, we can explore the implications of other's policies to inform our own input to policy consultations; we can also use it to develop our own policies; and it is a place for us to manage and refine our policy. The system puts core GIS processing and interrogation in the hands of 'non GIS specialists' as well as delivering information for consultation purposes. In this way, MaRS is assisting The Crown Estate to integrate existing and emerging marine policies into its planning activities.

The use of MaRS has highlighted some key benefits of the integration of planning and policy, including:

- The creation of a more certain environment for the planning of marine development.
- Mitigation of potential conflicts in use of the seabed at the planning stage.
- Flagging priorities for further policy development.
- A more structured and transparent approach to consultation with key stakeholders.

The Crown Estate's experience has also highlighted the importance of planning across jurisdictional boundaries. Marine management and planning around the world faces integration and boundary issues, whether these are at the very local level or at the ocean scale, such as the Large Marine Ecosystems programme (Carleton Ray and McCormick-Ray, 2004). The UK is not immune to this; there are numerous policies and legislative and statutory requirements at various scales but in some cases they do not relate or interlink with one another and in other instances the integration is far more effective. For 'all-use' marine management to work effectively, it is the coordination and integration of management that are essential and not where the geographical boundaries lie. The importance of public buy-in and involvement in the management and decision-making process is also paramount. It is essential, however, that focused guidelines regarding processes and objectives are created when dealing with such a wide range of stakeholders if any consultative process is to prove fruitful.

David Tudor and Tim Norman,
Marine Spatial Planning Team, The Crown Estate

References

Carleton Ray, G. and McCormick-Ray, J. (2004) *Coastal-Marine Conservation: Science and Policy*, Blackwell, Oxford

Cicin-Sain, B. and Knecht, R. W. (1998) *Integrated Coastal and Ocean Management: Concepts and Practices*, Island, Washington, DC

Foster, E., Haward, M. and Coffen-Smout, S. (2005) 'Implementing integrated oceans management: Australia's south east regional marine plan (SERMP) and Canada's eastern Scotian shelf integrated management (ESSIM) initiative', *Marine Policy*, vol 29, no 5, pp391–405

Shipman, B. and Stojanovic, T. (2007) 'Facts, fictions, and failures of integrated coastal zone management in Europe', *Coastal Management*, vol 35, nos 2–3, pp375–398

The pre-planning process of MSP organizes the structure, including the identification of a set of clear and measurable objectives. As with fisheries management, these objectives are generally (but not exclusively) based on sound understanding and data. To some extent, knowledge and data also contribute to defining the boundaries for analysis, for management, and for sources of influence – especially when MSP is based on bioregions (for example, the South West Marine Bioregion of Australia).

Step 5 of the UNESCO MSP guidelines involves defining and analysing the existing situation. The outputs delivered in this step include:

- inventory of maps of important biological and ecological areas; and
- inventory of maps of current human activities for the assessment of conflicts and compatibilities.

This is very much akin to marine habitat mapping, perhaps differing only in terms of aim (in other words, mapping distributions of species and human activities as well as habitats), utilizing common data sets (for example, ecological distributions, spatial patters of human activities, oceanographic data, and so on) and technologies as well as the current management frameworks that are variably integrated. An inventory of the current status of the coastal and marine environment brings together a wide range of baseline data, taking account of trends and developments. Here, data sets of habitat characteristics and indicators should be considered at a range of spatial and temporal levels. The inventory should aim to be as comprehensive as possible, including specific ecological characteristics that are particularly sensitive or ecologically important areas. In addition, the main pressures and threats should be identified, as well as the sectors that depend on a particular type of marine area, for example, Boyes et al (2007).

Directly reflecting the importance of marine data that underpin EA to fisheries management, data management is regarded as being as important as the data themselves (including documentation and metadata). Data should be handled and managed in the form of atlases, geodatabases and GIS. Geodatabases may be linked to data modelling handled within a spatial analysis framework.

Step 6 is the assessment of future conditions. This establishes 'where do we want to be?' as a consequence of management. The purpose here is to help envisage and plan how to create the desired future. Defining and analysing future conditions involves projecting current trends in the spatial and temporal needs of human uses, the new demands on open space, future scenarios for the planning area and selecting the preferred sea use scenario(s). In relation to temporal and spatial aspects of human need, future demands for food security (and sometimes other specialized uses of ocean space) are likely to limit the availability of open space in guiding management goals. The assessment of future conditions has direct parallels with the MP or MSE modelling approach, with an emphasis on the management model component. Fundamentally, this step is not based on data. However, the use scenarios may be overlain onto the present condition inventories to identify where conflicts are likely to arise. Similarly, the current knowledge base will be useful in identifying possible alternative futures driven by: (i) ecology and biodiversity; (ii) economy; or (iii) society and culture. Data also feed into these future considerations, such as the physical and environmental limits on given activities such as aggregate extraction and the location of offshore wind turbines. Preparing the MSP through zoning clearly has to identify where key areas may need protection, such as MPAs, nature conservation areas (fish, birds, reefs, wetlands).

Monitoring and evaluation performance of the MSP (Step 9) is a continuous management activity that uses systematic data collection of selected indicators to provide managers and stakeholders with evidence on the extent of progress towards the achievement of goals and objectives, thus directly paralleling the use of indicators in ecosystem-based fisheries management where extensive research has been undertaken on the efficacy, measurability, diagnostic qualities, and communicability of indicators of marine ecosystems. Monitoring and evaluation allows:

1. Assessment of the state of the system (for example, status of biodiversity); and
2. Measurement of the performance of management actions.

Monitoring objectives must be clear with respect to the questions that provide the basis for measurement. It is not only a data-gathering activity, but management, analysis, synthesis and interpretation – and is resourced accordingly. The key question being addressed through monitoring is that of separating human activity from natural variability, which is very difficult in a system that is inherently complex and variable. Lessons may be drawn in this respect from fisheries management research in the application of indicators, ecosystem models that incorporate extrinsic drivers and, indeed, the designation of MPAs as a means of assessing trends in fished versus non-fished areas. The use of spatialized indicators in zoning using a GIS platform (for example, Babcock et al, 2005) perhaps represents the first step in developing this process, again building on existing ecosystem-based fisheries management tools.

MSP monitoring includes modelling, laboratory and field research, time-series measurements, quality assurance, data analysis tools, synthesis and interpretation. In order to be able to detect trends and trajectories, baseline data on parameters and indicators are essential. As with the utility of indicators in tracking the delivery of an ecosystem approach to fisheries management, the indicators adopted in support of MSP may be quantitative or qualitative measures, the functions of which are simplification, quantification and communication, the characteristics of which are given in Table 5.7. Significantly, these indicators do not readily enable assessment of ecosystem function or status – conservation of which, of course, underlies all objectives of the ecosystem approach to management irrespective of its mode of delivery. Further development is therefore required in the development and application of ecosystem indicators with which to assess the management process.

In recognizing the parallels between the data and knowledge demands of an EA to fisheries management and MSP, it is apparent that both are founded on robust principles, sound theory and, in the case of fisheries management, well-established practice. While they do not explicitly draw on specific fisheries management experience, the MSP principles as outlined by Ehler and Douvere (2009) are closely aligned to the use of data time series and indicators in marine ecosystem management. Indeed, the demands of indicators in MSP are almost exactly the requirement of indicators of ecosystem function and health. Similarly, the modelling frameworks and approaches for fisheries management strategy scenario testing, the technologies available for determining the current inventory of resources and the platforms for handling and zoning spatially referenced data through time lend themselves well to MSP. The same also applies to the overarching need for the approach to be precautionary and adaptive.

Table 5.7 The characteristics of good indicators for MSP monitoring and evaluation

Readily measurable	On the timescales needed to support management, using existing instruments, monitoring programmes and available analytical tools.
Cost-effective	Monitoring resources are usually limited.
Concrete	Indicators that are directly observable and measurable are desirable because they are more readily interpretable and accepted by diverse stakeholder groups.
Interpretable	Indicators should reflect properties of concern to stakeholders; their meaning should be understood by as wide a range of stakeholders as possible.
Grounded in theory	Indicators should be based on well-accepted scientific theory, rather than on inadequately defined or poorly validated theoretical links.
Sensitive	Indicators should be sensitive to changes in the properties being monitored (for example, able to detect trends in the properties or impacts).
Responsive	Indicators should be able to measure the effects of management actions to provide rapid and reliable feedback on their performance and consequences.
Specific	Indicators should respond to the properties they are intended to measure rather than to other factors, in other words, it should be possible to distinguish the effects of other factors from the observed responses.

Source: Ehler and Douvere (2009)

Equally, one is tempted to identify the same challenges to the implementation of EA in both fisheries management and MSP in terms of data and knowledge deficiencies; incomplete consideration of spatial and temporal heterogeneity, uncertainty, and the full range of influences on resources (environmental, ecological, socio-economic); and imperfect knowledge exchange. However, where the two themes differ is perhaps in the timing and breadth of stakeholder engagement, earlier and wider in the case of MSP process, and a longer history of transdisciplinary knowledge sharing and establishing what works (and what does not) in guiding operational fisheries management.

The challenges ahead

In taking forward EA to managing the marine environment within an integrated MSP framework, it is clear that a lot of ground has already been covered, and indeed lessons learned, in the implementation of fisheries management, whether it is ecosystem-based or takes an EA. It would be a monumental oversight not to draw from the experiences of scientists, stakeholders, decision- and policy-makers in fisheries management where the current shortcomings and challenges that remain are likely to be also encountered in MSP. Indeed, the research challenges to MPA implementation already highlight where future work should be focused.

In terms of the use of data in underpinning current practice, data management initiatives are an important step towards developing common data protocols in support of tracking management progress – an essential requirement of adaptive implementation. Similarly, work on the use of indicators has revealed the critical importance of common, long-term data, perhaps requiring the modification or augmentation of current monitoring activities. In this data collection, there is often a tendency to collect easily acquired information rather than the most important. Similarly, socio-economic data are often ignored. Monitoring needs to be related to clear policy objectives in support of the decision-making process, and trust in data and interpretation needs to be built with stakeholders through dialogue and clear communication.

Models are still in their infancy, particularly when it comes to complex ecosystem models that aim to predict change as a result of internal dynamics or external forcing. Hence, they should be viewed as tools for identifying the direction of change in response to changing climate, environment and/or human action. Models are often viewed with caution by stakeholders and decision-makers because of their scientific focus (in other words, not related to policy objectives), their intangibility (poorly communicated to the non-specialist) and that many do not incorporate socio-economic data.

When it comes to the role of science in taking an EA, it is important to stress that policy decisions are rarely taken solely on the basis of long-term outcomes, short-term considerations are always in play and for several reasons often dominate. The short-term outcomes are generally less uncertain and are much more real to many participants in the decision-making process than longer term visions. In an EA to fisheries management, many of the decisions leading to preferred longer term ecological outcomes require short-term social or economic costs, and those bearing the costs may not feel they will be adequately compensated on either the longer term or shorter term. Also, the costs fall largely on people, whereas the benefits accrue to a large, complex and (to many people) abstract 'ecosystem'.

As a consequence, on all timescales, ecological perspectives are combined with economic and social perspectives in decision-making. Desirable ecological outcomes with high short-term social or economic costs are not going to be attractive to decision-makers, nor to the sectors of society that will bear the social and economic costs. To ensure all three aspects of the decision get appropriate weight, the ecological community needs to acknowledge the costs inherent in their preferred outcomes and collaborate with the social and economic science experts who can develop strategies for how those costs can best be addressed. At the same time, policy objectives need to be expressed as specifically as possible in order to direct the work of all the experts and make their findings more policy relevant. All participants in the governance processes need to embrace the concept of adaptive management based on a process of ongoing 'experimentation', using the best but inevitably imperfect understanding, tracked by monitoring and delivered in the context of periodic policy adjustment.

Progress on these challenges may be fastest at more local scales. Public and political engagement in marine management becomes more difficult at regional seas scales. As scale increases, the value of soundly constructed management solutions increases, but the solutions are difficult to implement. Similarly, the transdisciplinary perspectives in marine management become increasingly difficult to achieve due to the amplifying number of trade-offs to be resolved, increasingly fragmented expertise in the ecological and social sciences, barriers presented by sectoral administrative systems and the conflicting goals in overlapping international agreements. Experts need open minds, patience, and a diversity of collaborators and partners to overcome these challenges.

References

Agardy, T. (2000) 'Effects of fisheries on marine ecosystems: A conservationist's perspective', *ICES Journal of Marine Science*, vol 57, no 3, pp761–765

Agardy, T. (2003) 'An environmentalist's perspective on responsible fisheries: The need for holistic approaches', in M. Sinclair and G. Valdimarsson (eds) *Responsible Fisheries in the Marine Ecosystem*, FAO and CABI Publishing, Wallingford, Oxford, pp65–85

Andaloro, F., Baro, J., Coppola, S. R. and Koutrakis, E.T. (2002) 'Small scale fishery along the Mediterranean coast: New opportunities of development', *International Conference on Mediterranean Fisheries*, June 21–22, Naples

Anderson, C. N. K., Hsieh, C. S., Sandin, S. A., Hewitt, R., Hollowed, A., Beddington, J., May, R. M. and Sugihara, G. (2008) 'Why fishing magnifies fluctuations in fish abundance', *Nature*, vol 452, no 7189, pp835–839

Ault, J. S (1996) 'A fishery management system approach for Gulf of Mexico living resources', in P. J. Rubec and J. O'Hop (eds) *GIS Applications for Fisheries and Coastal Resources Management*, vol 43, Gulf States Marine Fisheries Commission, pp106–111

Auster, P. (1998) 'A conceptual model of the impacts of fishing gear on the integrity of fish habitats', *Conservation Biology*, vol 12, no 6, pp1198–1203

Babcock, E. A., Pikitch, E. K., McAllister, M. K., Apostolaki, P. and Santora, C. (2005) 'A perspective on the use of spatialized indicators for ecosystem-based fishery management through spatial zoning', *ICES Journal of Marine Science*, vol 62, no 3, pp469–476

Ball, I. R. (2000) *Mathematical Applications for Conservation Ecology: The Dynamics of Tree Hollows and the Design of Nature Reserves*, University of Adelaide

Ball, I. R., Possingham, H. P. and Watts, M. (2009) 'Marxan and relatives: Software for spatial conservation prioritisation', in A. Moilanen, K.A. Wilson, and H.P. Possingham (eds) *Spatial Conservation Prioritisation: Quantitative Methods and Computational Tools*, Oxford University Press, Oxford, pp185–195

Balmford, A., Bruner, A., Cooper, P., Costanza, R., Farber, S., Green, R. E., Jenkins, M., Jefferiss, P., Jessamy, V., Madden, J., Munro, K., Myers, N., Naeem, S., Paavola, J., Rayment, M., Rosendo, S., Roughgarden, J., Trumper, K. and Turner, R. K. (2002) 'Economic reasons for conserving wild nature', *Science*, vol 297, no 5583, pp950–953

Barange, M. (2005) *Science for Sustainable Marine Bioresources*, report for the Natural Environment Research Council, Defra and the Scottish Executive for Environment and Rural Affairs, Plymouth Marine Laboratory, Plymouth

Baretta, J. W., Ebenhöh, W. and Ruardij, P. (1995) 'The European Regional Seas Ecosystem Model, a complex marine ecosystem model', *Netherlands Journal of Sea Research*, vol 33, nos 3–4, pp233–246.

Battista, T. A. and Monaco, M. E. (2004) 'Geographic information systems applications in coastal marine fisheries', in W. L. Fisher and F. J. Rahel (eds) *Geographic Information Systems in Fisheries*, American Fishery Society, Bethesda, Maryland, pp189–208

Berkson, J. M., Kline, L. and Orth, D. J. (eds) (2002) *Incorporating Uncertainty into Fishery Models*, Symposium 27, American Fisheries Society, Bethesda, Maryland

Beverton, R. J. H. and Holt, S. J. (1957) *On the Dynamics of Exploited Fish Populations*, Chapman and Hall, London

Bianchi, G., Gislason, H., Graham, K., Hill, L., Jin, X., Koranteng, K., Manickchand-Heileman, S., Payá, I., Sainsbury, K., Sanchez, F. and Zwanenburg, K. (2000) 'Impact of fishing on size composition and diversity of demersal fish communities', *ICES Journal of Marine Science*, vol 57, no 3, pp558–571

Blackford, J. C. and Radford, P. J. (1995) 'A structure and methodology for marine ecosystem modelling', *Netherlands Journal of Sea Research*, vol 33, nos 3–4, pp247–260

Blackford, J. C., Allen, J. I. and Gilbert, F. J. (2004) 'Ecosystem dynamics at six contrasting sites: A generic modelling study', *Journal of Marine Systems*, vol 52, nos 1–4, pp191–215

Blanchard, J. L., Coll, M., Trenkel, V. M., Vergnon, R., Yemane, D., Jouffre, D., Link, J. S. and Shin, Y.-J. (2010) 'Trend analysis of indicators: A comparision of recent changes in the status of marine ecosystems around the world', *ICES Journal of Marine Science*, vol 67, no 4, pp732–744

Boyes, S. J., Elliott, M., Thomson, S. M., Atkins, S. and Gilliland, P. (2007) 'A proposed multiple-use zoning scheme for the Irish Sea: An interpretation of current legislation through the use of GIS-based zoning approaches and effectiveness for the protection of nature conservation interests', *Marine Policy*, vol 31, no 3, pp287–298

Butterworth, D. S. and Punt, A. E. (1999) 'Experiences in the evaluation and implementation of management procedures', *ICES Journal of Marine Science*, vol 56, no 6, pp985–998

Butterworth, D. S. and Punt, A. E. (2003) 'The role of harvest control laws, risk and uncertainty and the precautionary approach in ecosystem-based management', in M. J. Sinclair and H. Valdimarsson (eds) *Responsible Fisheries in the Marine Ecosystem*, FAO, Rome, pp311–319

Butterworth, D. S. and Rademeyer, R. A. (2005) 'Sustainable management initiatives for the southern African hake fisheries over recent years', *Bulletin of Marine Science*, vol 76, no 2, pp287–320

Cairns, J., McCormick, P. V. and Niederlehner, B. R. (1993) 'A proposed framework for developing indicators of ecosystem health', *Hydrobiologia*, vol 263, pp1–44

Casini, M., Lövgren, J., Hjelm, J., Cardinale, M., Molinero, J.-C. and Kornilovs, G. (2008) 'Multi-level trophic cascades in a heavily exploited open marine ecosystem', *Proceedings of the Royal Society of London B: Biological Sciences*, vol 275, pp1793–1801

CBD (Convention on Biological Diversity) (2000) *Decision V/6 adopted by the Conference of Parties to the Convention on Biological Diversity at its Fifth Meeting*, Nairobi, Kenya

Charbonnier, A. and Caddy, J. F. (1986) *Report of GFCM on the Methods of Evaluating Small Scale Fisheries in the Western Mediterranean. Sete, France 13–16 May 1986*, FAO Fisheries Report 362

Chifflet, M., Andersen, V., Prieur, L. and Dekeyser, I. (2001) 'One-dimensional model of short-term dynamics of the pelagic ecosystem in the NW Mediterranean Sea: Effects of wind events', *Journal of Marine Systems*, vol 30, nos 1–2, pp89–114

Christensen, V. and Walters, C. J. (2004) 'Ecopath with Ecosim: Methods, capabilities and limitations', *Ecological Modelling*, vol 172, nos 2–5, pp109–139

Christensen, V., Guenette, S., Heymans, S., Walters, C. J., Watson, R., Zeller, D. and Pauly, D. (2001) 'Estimating fish abundance of the North Atlantic, 1950–2000', in

S. Guenette, V. Christensen and D. Pauly (eds) *Fisheries Impacts on North Atlantic Ecosystems: Models and Analyses*, Fisheries Centre Research Report 9, University of British Columbia, Vancouver, pp1–25

Chuenpagdee, R., Morgan, L. E., Maxwell, S. M., Norse, E. A. and Pauly, D. (2003) 'Shifting gears: Assessing collateral impacts of fishing methods in US waters', *Frontiers in Ecology and the Environ*ment, vol 1, no 10, pp517–524

Clark, J.R. (1992) *Integrated Management of Coastal Zones*, FAO, Fisheries Technical Paper 327

Cogan, C. B., Todd, B. J., Lawton, P. and Noji, T. T. (2009) 'The role of marine habitat mapping in ecosystem-based management', *ICES Journal of Marine Science*, vol 66, no 9, pp2033–2042

Collet, S. (1999) 'Regionalisation and eco-development: Which pathway for artisanal fishers?', in D. Symes (ed) *Europe's Southern Waters: Management Issues and Practice*, Blackwell Science, Oxford, pp42–52

Collie, J. S., Escanero, G. A. and Valentine, P. C. (1997). 'Effects of bottom fishing on the benthic megafauna of Georges Bank', *Marine Ecology Progress Series*, vol 155, pp159–172

Connor, D., Gilliland, P., Golding N., Robinson, P., Todd, D. and Verling, E. (2006) *UKSeaMap: The Mapping of Seabed and Water Column Features of UK Seas*, Joint Nature Conservation Committee, Peterborough

Constable, A. J., de la Mare, W. K., Agnew, D. J., Everson, I. and Miller, D. (2000) 'Managing fisheries to conserve the Antarctic marine ecosystem: Practical implementation of the Convention on the Conservation of Antarctic Marine Living Resources (CCAMLR)', *ICES Journal of Marine Science*, vol 57, no 3, pp778–791

Cooke, J. G. (1999) 'Improvement of fishery-management advice through simulation testing of harvest algorithms', *ICES Journal of Marine Science*, vol 56, no 6, pp797–810

Crispi, G., Mosetti, R., Solidoro, C. and Crise, A. (2001) 'Nutrients cycling in Mediterranean basins: The role of the biological pump in the trophic regime', *Ecological Modelling*, vol 138, nos 1–3, pp101–114

Cury, P. M. and Christensen, V. (2005) 'Quantitative ecosystem Indicators for fisheries management', *ICES Journal of Marine Science*, vol 62, no 3, pp307–310

Cury, P. M., Shannon, L. and Shin, Y.-J. (2003) 'The functioning of marine ecosystems: A fisheries perspective', in M. Sinclair and G. Valdimarsson (eds) *Responsible Fisheries in the Marine Ecosystem*, FAO and CABI Publishing, Wallingford, Oxford, pp103–123

Cury, P. M., Mullon, C., Garcia, S. M. and Shannon, L. J. (2005a) 'Viability theory for an ecosystem approach to fisheries', *ICES Journal of Marine Science*, vol 62, no 3, pp577–584

Cury, P. M., Shannon, L. J., Roux, J.-P., Daskalov, G. M., Jarre, A., Moloney, C. L. and Pauly, D. (2005b) 'Trophodynamic indicators for an ecosystem approach to fisheries', *ICES Journal of Marine Science*, vol 62, no 3, pp430–442

Dayton, P. K., Thrush, S. F., Agardy, M. T. and Hofman, R. J. (1995) 'Environmental effects of marine fishing', *Aquatic Conservation: Marine and Freshwater Ecosystems*, vol 5, no 3, pp205–232

De Groot, R. (1992) *Functions of Nature*, Wolters-Noordhoff, Amsterdam

De la Mare, W. K. (1998) 'Tidier fisheries management requires a new MOP (management-orienated paradigm)', *Reviews in Fish Biology and Fisheries*, vol 8, no 3, pp349–356

Defra (Department for Environment, Food and Rural Affairs) (2002) *General Fisheries Technical Conservation Rules*, http://Scotland.gov.uk/library5/fisheries.gftcr.pdf

Defra (2005) *Charting Progress: An Integrated Assessment of the State of the Seas*, Defra, London

Dickson, R. R., Kelly, P. M., Colebrook, J. M., Wooster, W. S. and Cushing, D. H. (1988) 'North winds and production in the eastern North Atlantic', *Journal of Plankton Research*, vol 10, no 1, pp151–169

Dinmore, T. A., Duplisea, D. E., Rackham, B. D., Maxwell, D. L. and Jennings, S. (2003) 'Impact of a large-scale area closure on patterns of fishing disturbance and the consequences for benthic communities', *ICES Journal of Marine Science*, vol 60, no 2, pp371–380

Douvere, F. and Ehler, C. N. (2007) 'International Workshop on Marine Spatial Planning, UNESCO, Paris, 8–10 November 2006: A summary', *Marine Policy*, vol 31, no 4, pp582–583

Douvere, F., Maes, F., Vanhulle, A. and Schrijvers, J. (2007) 'The role of marine spatial planning in sea use management: The Belgian case', *Marine Policy*, vol 31, no 2, pp182–191

Drouineau, H., Mahévas, S., Pelletier, D. and Beliaeff, B. (2006) 'Assessing the impact of different management options using ISIS-Fish: The French *Merluccius merluccius-Nethrops norvegicus* mixed fishery of the Bay of Biscay', *Aquatic Living Resources*, vol 19, no 1, pp15–29

Duplisea, D. E., Jennings, S., Warr, K. J. and Dinmore, T. A. (2002) 'A size-based model of the impacts of bottom trawling on benthic community structure', *Canadian Journal of Fisheries and Aquatic Sciences*, vol 59, no 11, pp1785–1795

Eastwood, P. D., Mills, C. M., Aldridge, J. N., Houghton, C. A. and Rogers, S. I. (2007) 'Human activities in UK offshore waters: An assessment of direct, physical pressure on the seabed', *ICES Journal of Marine Science*, vol 64, no 3, pp453–463

Ehler, C. and Douvere, F. (2009) *Marine Spatial Planning: A Step-by-step Approach Toward Ecosystem-based Management*, Intergovernmental Oceanographic Commission and Man and the Biosphere Programme, IOC Manual and Guides No 53, ICAM Dossier No 6, UNESCO, Paris

Elith, J. and Graham, C. H. (2009) 'Do they? How do they? WHY do they differ? On finding reasons for differing performances of species distribution models', *Ecography*, vol 32, no 1, pp66–77

Embling, C. B., Gillibrand, P. A., Gordon, J., Shrimpton, J., Stevick, P. T. and Hammond, P. S. (2010) 'Using habitat models to identify suitable sites for marine protected areas for harbour porpoises (*Phocoena phocoena*)', *Biological Conservation*, vol 143, no 2, pp267–279

FAO (Food and Agriculture Organization) (1996) *Precautionary Approach to Fisheries: Guidelines on the Precautionary Approach to Capture Fisheries and Species Introductions*, FAO Fisheries Technical Paper 350-1

FAO (2002) *Report of the Reykjavik Conference on Responsible Fishing*, FAO Fisheries Report 658

FAO (2003) *The Ecosystem Approach to Fisheries*, FAO Technical Guidelines for Responsible Fisheries

FAO (2009) *The State of World Fisheries and Aquaculture 2008*, FAO, Rome, available at www.fao.org/fishery/sofia/en

Farruggio, H. (1989) 'Artisanal et pêche en Méditerranée évolution et état thé la recherché', *La recherche Face a la Pêche Artisanal Symposium International*, ORSTOM – IFREMER, Montpellier, 3–5 July 1989

Fielding, A. H. and Bell, J. F. (1997) 'A review of methods for the assessment of prediction errors in conservation presence/absence models', *Environmental Conservation*, vol 24, pp38–49

Fisher, W. S., Jackson, L. E., Suter, G. W. and Bertram, P. (2001) 'Indicators for human and ecological risk assessment: A US EPA perspective', *Human and Ecological Risk Assessment*, vol 7, no 5, pp961–970

Fogarty, M. J. and Murawski, S. A. (1998) 'Large-scale disturbance and the structure of marine systems: Fishery impacts on Georges Bank', *Ecological Applications*, vol 8, ppS6–S22

Fréon, P., Drapeau, L., David, J. H. M., Fernández Moreno, A., Leslie, R. W., Oosthuizen, W. H., Shannon, L. J. and van der Lingen, C. D. (2005) 'Spatialized ecosystem indicators in the southern Benguela', *ICES Journal of Marine Science*, vol 62, no 3, pp459 468

Frid, C. L. J., Paramor, O. A. L. and Scott, C. L. (2005) 'Ecosystem-based fisheries management: Progress in the NE Atlantic', *Marine Policy*, vol 29, no 5, pp461–469

Frid, C. J. L., Paramor, O. A. L. and Scott, C. L. (2006) 'Ecosystem-based management of fisheries: Is science limiting?', *ICES Journal of Marine Science*, vol 63, no 9, pp1567–1572

Garcia, S. M. (2000) 'The FAO definition of sustainable development and the Code of Conduct for Responsible Fisheries: An analysis of the related principles, criteria and indicators', *Marine & Freshwater Research*, vol 51, no 5, pp535–541

Garcia, S. M. (2005) 'Fishery science and decision-making: Dire straights to sustainability', *Bulletin of Marine Science*, vol 76, no 2, pp171–196

Garcia, S. M. and Staples, D. J. (2000) 'Sustainability reference systems and indicators for responsible marine capture fisheries: A review of concepts and elements for a set of guidelines', *Marine & Freshwater Research*, vol 51, no 5, pp385–426

Garibaldi, L. and Caddy, S. M. (1998) 'Biogeographic characterization of Mediterranean and Black Seas faunal provinces using GIS procedures', *Ocean & Coastal Management*, vol 39, no 3, pp211–227

Gerber, L. R., Botsford, L. W., Hastings, A., Possingham, H. P., Gaines, S. D., Palumbi, S. R. and Andelman, S. (2003) 'Population models for marine reserve design: A retrospective and prospective synthesis', *Ecological Applications*, vol 13, no 1, ppS47–S64

Gerber, L. R., Wielgus, J. and Sala, E. (2007) 'A decision framework for the adaptive management of an exploited species with implications for marine reserves', *Conservation Biology*, vol 21, no 6, pp1594–1602

Greenstreet, S. P. R. and Rogers, S. I. (2006) 'Indicators of the health of the North Sea fish community: Identifying reference levels for an ecosystem approach to management', *ICES Journal of Marine Science*, vol 63, no 4, pp573–593

Guisan, A. and Zimmerman, N. E. (2000) 'Predictive habitat distribution models in ecology', *Ecological Modelling*, vol 135, nos 2–3, pp147–186

Haedrich, R. L., Devine, J. A. and Kendall, V. J. (2008) 'Predictors of species richness in the deep-benthic fauna of the northern Gulf of Mexico', *Deep-Sea Research II*, vol 55, pp2650–2656

Hall, S J. (1994) 'Physical disturbance and marine benthic communities: Life in unconsolidated sediments', *Oceanography and Marine Biology: An Annual Review*, vol 32, pp179–239

Hall, S. J. (1999) 'Managing fisheries within ecosystems: Can the role of reference points be expanded?', *Aquatic Conservation: Marine and Freshwater Ecosystems*, vol 9, no 6, pp579–583

Hardy, A. (1956) *The Open Sea*, Collins, London

Hazin, H. and Erzini, K. (2008) 'Assessing swordfish distribution in the South Atlantic from spatial predictions', *Fisheries Research*, vol 90, nos 1–3, pp45–55

Hiddink, J. G., Hutton, T., Jennings, S. and Kaiser, M. J. (2006) 'Predicting the effects of area closures and fishing effort restrictions on the production, biomass, and species richness of benthic invertebrate communities', *ICES Journal of Marine Science*, vol 63, no 5, pp822–830

Higgason, K. D. and Brown, M. (2009) 'Local solutions to manage the effects of global climate change on a marine ecosystem: A process guide for marine resource managers', *ICES Journal of Marine Science*, vol 66, no 7, pp1640–1646

Hilborn, R. (1985) 'Fleet dynamics and individual variation: Why some people catch more fish than others', *Canadian Journal of Fisheries and Aquatic Sciences*, vol 42, no 1, pp2–13

Hoffmann, E. and Pérez-Ruzafa, A. (2009) 'Marine Protected Areas as a tool for fishery management and ecosystem conservation: An introduction', *ICES Journal of Marine Science*, vol 66, no 1, pp1–5

Holland, D. S. (2003) 'Integrating spatial management measures into traditional fishery management systems: The case of the Georges Bank multispecies groundfish fishery', *ICES Journal of Marine Science*, vol 60, no 5, pp915–929

Hollowed, A. B., Bax, N., Beamish, R., Collie, J., Fogarty, M., Livingston, P., Pope, J. and Rice, J. C. (2000) 'Are multispecies models an improvement on single-species models for measuring fishing impacts on marine ecosystems?', *ICES Journal of Marine Science*, vol 57, no 3, pp707–719

Horwood, J. W., Nichols, J. H. and Milligan, S. (1998) 'Evaluation of closed areas for fish stock conservation', *Journal of Applied Ecology*, vol 35, no 6, pp893–903

Hoydal, K. (2007) 'Viewpoint: The interface between scientific advice and fisheries management', *ICES Journal of Marine Science*, vol 64, no 4, pp846–850

Hsieh, C. H., Reiss, C. S., Hunter, J. R., Beddington, J. R., May, R. M. and Sugihara, G. (2006) 'Fishing elevates variability in the abundance of exploited species', *Nature*, vol 443, no 7114, pp859–862

Hutchings, J. A and Myers, R. A. (1994) 'What can be learned from the collapse of a renewable resource? Atlantic cod, *Gadus morhua*, of Newfoundland and Labrador', *Canadian Journal of Fisheries and Aquatic Sciences*, vol 51, no 9, pp2126–2146

ICES (International Council for the Exploration of the Sea) (2000) *Report of the Working Group on the Ecosystem Effects of Fishing Activities*, ICES, Copenhagen

ICES (2005) *Report of the ICES Advisory Committee on Fishery Management*, Advisory Committee on the Marine Environment, and Advisory Committee on Ecosystems, ICES Copenhagen

ICES (2006) *Report of the Working Group on the Ecosystem Effects of Fishing Activity*, ICES, Copenhagen

Jennings, S. (2005) 'Indicators to support an ecosystem approach to fisheries', *Fish and Fisheries*, vol 6, no 3, pp212–232

Jennings, S. (2009) 'The role of marine protected areas in environmental management', *ICES Journal of Marine Science*, vol 66, no 1, pp16–21

Jennings, S. and Dulvy, N. K. (2005) 'Reference points and reference directions for size-based indicators of community structure', *ICES Journal of Marine Science*, vol 62, no 3, pp397–404

Jennings, S. and Kaiser, M. J. (1998) 'The effects of fishing on marine ecosystems', *Advances in Marine Biology*, vol 34, pp201–352

Jennings, S., Greenstreet, S. P. R. and Reynolds, J. D. (1999) 'Structural change in an exploited fish community: A consequence of differential fishing effects on species with contrasting life histories', *Journal of Animal Ecology*, vol 68, no 3, pp617–627

Jennings, S., Kaiser, M. J. and Reynolds, J. D. (2001) *Marine Fisheries Ecology*, Blackwell Science, Oxford

Kell, L. T., Pilling, G. M., Kirkwood, G. P., Pastoors, M. A., Mesnil, B., Korsbrekke, K., Abaunza, P., Aps, R., Biseau, A., Kunzlik, P., Needle, C. L., Roel, B. A. and Ulrich, C. (2006) 'An evaluation of multi-annual management strategies for ICES roundfish stocks', *ICES Journal of Marine Science*, vol 63, no 1, pp12–24

Kemp, Z. and Meaden, G. (2002) 'Visualization for fisheries management from a spatiotemporal perspective', *ICES Journal of Marine Science*, vol 59, no 1, pp190–202

Koslow, J. A. (2009) 'The role of acoustics in ecosystem-based fishery management', *ICES Journal of Marine Science*, vol 66, no 6, pp966–973

Kraus, G., Pelletier, D., Dubreuil, J., Möllmann, C., Hinrichsen, H.-H., Bastardie, F., Vermard, Y. and Mahévas, S. (2009) 'A model-based evaluation of Marine Protected Areas: The example of eastern Baltic cod (*Gadus morhua callarias L.*)', *ICES Journal of Marine Science*, vol 66, no 1, pp109–121

Krause, M. and Trahms, J. (1983) *Zooplankton Dynamics During FLEX '76*, Springer, Berlin

Leonard, J. and Maynou, F. (2003) 'Fish stocks assessments in the Mediterranean: State of the art', *Scientia Marina*, vol 67, pp37–49

Lindholm, J. B., Auster, P. J., Ruth, M. and Kaufman. L. (2001) 'Modeling the effects of fishing and implications for the design of marine protected areas: Juvenile fish responses to variations in seafloor habitat', *Conservation Biology*, vol 15, no 2, pp424–437

Mahévas, S. and Pelletier, D. (2004) 'ISIS-Fish, a generic and spatially explicit simulation tool for evaluating the impact of management measures on fisheries dynamics', *Ecological Modelling*, vol 171, nos 1–2, pp65–84

Makris, N. C., Ratilal, P., Symonds, D. T., Jagannathan, S., Lee, S. and Nero, R. W. (2006) 'Fish population and behaviour revealed by instantaneous continental shelf-scale imaging', *Science*, vol 311, no 5761, pp660–663

Mangi, S. C., Hattam, C., Rodwell, L. D., Rees, S. and Stehfest, K. (2009) *Initial Report on Socio-economic Costs of Closing Lyme Bay to Scallop Dredging and Heavy Trawling Gear*, report to Defra, June 2009

Marshall, C. E., Glegg, G., Howell, K. L., Embling, C. B., Langston, B. and Stevens, T. (2010) 'Using predictive distribution modelling to inform monitoring of benthic recovery in a temperate marine protected area: A case study using a shallow water gorgonian', in review

Mayer, F. L. and Ellersieck, M. R. (1986) *Manual of Acute Toxicity: Interpretation and Data Base for 410 Chemicals and 66 Species of Freshwater Animals*, Resource Publication 160, US Fish and Wildlife Service, Washington, DC

MEA (Millennium Ecosystem Assessment) (2005) *Ecosystems and Human Well-being: Synthesis*, Island, Washington, DC

Megrey, B. and Hinckley, S. (2001) 'Effect of turbulence on the feeding of larval fishes: A sensitivity analysis using an individual-based model', *ICES Journal of Marine Science*, vol 58, no 5, pp1015–1029

Murawski, S. (2000) 'Definitions of overfishing from an ecosystem perspective', *ICES Journal of Marine Science*, vol 57, no 3, pp649–658

Murawski, S. (2007) 'Ten myths concerning ecosystem approaches to marine resource management', *Marine Policy*, vol 31, no 6, pp681–690

Murawski, S. A., Brown, R., Lai, H. L., Rago, P. R. and Hendrickson, L. (2000) 'Large-scale closed areas as a fishery management tool in temperate marine systems: The Georges Bank experience', *Bulletin of Marine Science,* vol 66, no 3, pp775–798

Nicholson, M. D. and Jennings, S. (2004) 'Testing candidate indicators to support ecosystem-based management: The power of monitoring surveys to detect temporal trends in fish community metrics', *ICES Journal of Marine Science*, vol 61, no 1, pp35–42

Noji, T. T., Fromm, S., Vitaliano, J. and Smith, K. (2008) *Habitat Suitability Modelling Using the Kostylev Approach as an Indicator of Distribution of Benthic Invertebrates*, ICES Document CM 2008/G:15

NRC (National Research Council) (2000) *Ecological Indicators for the Nation*, National Academy Press, Washington, DC

OSPAR (1992) *Convention for the Protection of the Marine Environment of the North-east Atlantic, Annex V*, The Protection and Conservation of the Ecosystems and Biological Diversity of the Maritime Areas, www.ospar.org

Panzeri, M. and Morris, K. (2000) 'The integration of spatial and temporal data for consortia-based initiatives: The use of a 4D GIS', in *Oceanology 2000*, pp337–347

Pascoe, S. (2006) 'Economics, fisheries, and the marine environment', *ICES Journal of Marine Science*, vol 63, no 1, pp1–3

Patterson, K. R., Cook, R. M., Darby, C. D., Gavaris, S., Kell, L., Lewy, P., Mesnil, B., Punt, A., Restrepo, V., Skagen, D. W. and Stefánsson, G. (2001) 'Estimating uncertainty in fish stock assessment and forecasting', *Fish and Fisheries*, vol 2, no 2, pp125–157

Pattison, D., dos Reis, D. and Smilie, H. (2004) *An Inventory of GIS-Based Decision-Support Tools for MPAs*, prepared by the National MPA Center in cooperation with the National Oceanic and Atmospheric Administration Coastal Services Center

Pauly, D. (2006) 'Major trends in small-scale marine fisheries, with emphasis on developing countries, and some implications for the social sciences', *MAST*, vol 4, no 2, pp7–22

Pauly, D., Christensen, V. and Walters, C. (2000) 'Ecopath, Ecosim and Ecospace as tools for evaluating ecosystem impact of fisheries', *ICES Journal of Marine Sciences*, vol 57, no 3, pp697–706

Pauly, D., Christensen, V., Dalsgaard, J., Froese, R. and Torres, F. Jr. (1998) 'Fishing down marine food webs', *Science*, vol 279, pp860–863

Pauly, D., Christensen, V., Guénette, S., Pitcher, T. J., Sumalia, U.R., Walters, C. J., Watson, R. and Zeller, D. (2002) 'Towards sustainability in world fisheries', *Nature*, vol 418, no 6898, pp689–695

Peckett, F. J., Glegg, G. A. and Rodwell, L. D. (2010) *Assessing the Quality of Data Required to Protect Marine Biodiversity*, working paper, University of Plymouth

Pelletier, D. and Mahévas, S. (2005) 'Spatially explicit fisheries simulation models for policy evaluation', *Fish and Fisheries*, vol 6, no 4, pp307–349

Petihakis, G., Smith, C. J., Triantafyllou, G., Sourlantzis, G., Papadopoulou, K.-N., Pollani, A. and Korres, G. (2007) 'Scenario testing of fisheries management strategies using a high resolution ERSEM-POM ecosystem model', *ICES Journal of Marine Science*, vol 64, no 9, pp1627–1640

Piet, G. J., Jansen, H. M. and Rochet, M.-J. (2008) 'Evaluating potential indicators for an ecosystem approach to fishery management in European waters', *ICES Journal of Marine Science*, vol 65, no 8, pp1449–1455

Pikitch, E., Santora, C., Babcock, E. A., Bakun, A., Bonfil, R., Conover, D. O., Dayton, P., Doukakis, P., Fluharty, D., Heneman, B., Houde, E. D., Link, J., Livingston, P. A., Mangel, M., McAllister, M. K., Pope, J. and Sainsbury, K. J. (2004) 'Ecosystem-based fishery management', *Science*, vol 305, no 5682, pp346–347

Pinnegar, J. K., Blanchard, J. L., Mackinson, S., Scott, R. D and Duplisea, D. E. (2005) 'Aggregation and removal of weak-links in food-web models: System stability and recovery from disturbance', *Ecological Modelling*, vol 184, nos 2–4, pp229–248

Pitcher, T. J. and Preikshot, D. (2001) 'RAPFISH: A rapid appraisal technique to evaluate the sustainability status of fisheries', *Fisheries Research*, vol 49, no 3, pp255–270

Plagányi, É. E. and Butterworth, D. S. (2004) 'A critical look at the potential of Ecopath with Ecosim to assist in practical fisheries management', *African Journal of Marine Science*, vol 26, pp261–287

Polovina, J. J. and Howell, E. A. (2005) 'Ecosystem indicators derived from satellite remotely sensed oceanographic data for the North Pacific', *ICES Journal of Marine Science*, vol 62, no 3, pp319–327

Ponte, S., Raakjaer, J. and Campling, L. (2007) 'Swimming upstream: Market access for African fish exports in the context of WTO and EU negotiations and regulation', *Development Policy Review*, vol 25, pp113–138

Pope, J. G., Rice, J. C., Daan, N., Jennings, S. and Gislason, H. (2006) 'Modelling an exploited marine fish community with 15 parameters: Results from a simple size-based model', *ICES Journal of Marine Science*, vol 63, no 6, pp1029–1044

Possingham, H., Ball, I. and Andelman, S. (2002) 'Mathematical methods for identifying representative reserve networks', in S. Ferson, and M. Burgman (eds) *Quantitative*

Methods for Conservation Biology, Springer-Verlag, New York, pp291–305

Power, M. E., Tilman, D., Estes, J. A., Menge, B. A., Bond, W. J., Mills, L. S., Daily, G., Castilla, J. C., Lubchenco, J. and Paine, R. T. (1996) 'Challenges in the quest for keystones', *BioScience*, vol 48, no 8, pp609–620

Powles, H., Bradford, M. J., Bradford, R. G., Doubleday, W. G., Innes, S. and Levings, C. D. (2000) 'Assessing and protecting endangered marine species', *ICES Journal of Marine Science*, vol 57, no 3, pp669–676

Punt, A. E. and Hilborn, R. (1997) 'Fisheries stock assessment and decision analysis: The Bayesian approach', *Reviews in Fish Biology and Fisheries*, vol 7, no 1, pp35–63

Punt, A. E., Smith, A. D. M. and Cui, G. (2001) 'Review of progress in the introduction of management strategy evaluation (MSE) approaches in Australia's South East Fishery', *Marine & Freshwater Research*, vol 52, no 4, pp719–726

Rademeyer, R. A., Plagányi, É. E. and Butterworth, D. S. (2007) 'Tips and tricks in designing management procedures', *ICES Journal of Marine Science*, vol 64, no 4, pp618–625

Rees, H. L., Hyland, J. L., Hylland, K., Mercer Clarke, C. S. L., Roff, J. C. and Ware, S. (2008) 'Environmental indicators: Utility in meeting regulatory needs. An overview', *ICES Journal of Marine Science*, vol 65, no 8, pp1381–1386

Rees, S. E., Rodwell, L. D., Attrill, M. J., Austen, M. C and Mangi, S. C. (2010a) 'The value of marine biodiversity to the leisure and recreation industry and its application to marine spatial planning', *Marine Policy*, vol 34, no 5, pp868–875

Rees, S. E., Attrill, M. J., Austen, M. C., Mangi, S. C., Richards, J. P. and Rodwell, L. D. (2010b) 'Is there a win-win scenario for marine nature conservation? A case study of Lyme Bay, England', *Ocean & Coastal Management*, vol 53, no 3, pp135–145

Rice, J. and Ridgeway, L. (2009) 'Conservation of biodiversity in fisheries management', in R. Q. Grafton, R. Hilborn, D. Squires, M. Tait and M. Williams (eds) *Handbook of Marine Fisheries Conservation and Management*, Oxford University Press, Oxford, pp139–149

Rice, J. C. (1999) 'How complex should operational ecosystem objectives be?', CM Z:07, ICES, Copenhagen

Rice, J. C. (2000) 'Evaluating fishery impacts using metrics of community structure', *ICES Journal of Marine Science*, vol 57, no 3, pp682–688

Rice, J. C. (2005) 'Every which way but up: The sad story of Atlantic groundfish, featuring Northern cod and North Sea cod', *Bulletin of Marine Science*, vol 78, no 3, pp429–465

Rice, J. C. (2009) 'A generalisation of the three-stage model for advice using a Precautionary Approach in fisheries, to apply broadly to ecosystem properties and pressures', *ICES Journal of Marine Science*, vol 66, no 3, pp433–444

Rice, J. C. and Legacè, È. (2007) 'When control rules collide: A comparison of fisheries management reference points and IUCN criteria for assessing risk of extinction', *ICES Journal of Marine Science*, vol 64, no 4, pp718–722

Rice, J. C. and Rivard, D. (2007) 'The dual role of indicators in optimal fisheries management strategies', *ICES Journal of Marine Science*, vol 64, no 4, pp775–778

Rice, J. C. and Rochet, M.-J. (2005) 'A framework for selecting a suite of indicators for fisheries management', *ICES Journal of Marine Science*, vol 62, no 3, pp516–527

Rice, J. C., Trujillo, V., Jennings, S., Hylland, K., Hagstrom, O., Astudillo, A. and Jensen, J. N. (2005) *Guidance on the Application of the Ecosystem Approach to Management of Human Activities in the European Marine Environment*, ICES Cooperative Research Report 273, ICES, Copenhagen

Richards, L. J. and Maguire, J.-J. (1998) 'Recent international agreements and the precautionary approach: New directions for fisheries management science', *Canadian Journal of Fisheries and Aquatic Sciences*, vol 55, no 6, pp1545–1552

Richardson, K., Nielsen, T. G., Pedersen, F. B., Heilmann, J. P., Løekkegaard, B. and Kaas, H. (1998) 'Spatial heterogeneity in the structure of the planktonic food web in the North Sea', *Marine Ecology Progress Series*, vol 168, pp197–211

Ricker, W. E. (1975) 'Computation and interpretation of biological statistics of fish populations', *Bulletin of the Fisheries Research Board of Canada*, vol 191, pp1–382

Ridgeway, L. R. (2009) 'Governance beyond areas of national jurisdiction: Linkages to sectional management', *Oceanis: Towards a New Governance of High Seas Biodiversity*, Institute for Sustainable Development and International Relations, Paris

Rijnsdorp, A. D., Buys, A. M., Storbeck, F. and Visser, E. G. (1998) 'Micro-scale distribution of beam trawl effort in the southern North Sea between 1993 and 1996 in relation to the trawling frequency of the sea bed and the impact on benthic organisms', *ICES Journal of Marine Science*, vol 55, no 3, pp403–419

Roberts, C. M., Bohnsack, J. A., Gell, F., Hawkins, J. P. and Goodridge, R. (2001) 'Effects of marine reserves on adjacent fisheries', *Science*, vol 294, no 5548, pp1920–1923

Roberts, C. M., Gell, F. R. and Hawkins, J. P. (2003) *Protecting Nationally Important Marine Areas in the Irish Sea Pilot Project Region*, report to the Joint Nature Conservation Committee

Robinson, L. A. and Frid, C. L. J. (2003) 'Dynamic ecosystem models and the evaluation of ecosystem effects of fishing: Can we make meaningful predictions?', *Aquatic Conservation: Marine and Freshwater Ecosystems*, vol 13, no 1, pp5–20

Rochet, M.-J. and Rice, J. C. (2005) 'Do explicit criteria help in selecting indicators for ecosystem-based fisheries management?', *ICES Journal of Marine Science*, vol 62, no 3, pp528–539

Rochet, M.-J. and Rice, J. C. (2009) 'Simulation-based management strategy evaluation: Ignorance disguised as mathematics?', *ICES Journal of Marine Science*, vol 66, no 4, pp754–762

Sainsbury, K. J., Punt, A. E. and Smith, A. D. M. (2000) 'Design of operational management strategies for achieving fishery ecosystem objectives', *ICES Journal of Marine Science*, vol 57, no 3, pp731–741

Sala, E., Aburto-Oropeza, O., Paredes, G., Parra, I., Barrera, J. C. and Dayton, P. K. (2002) 'A general model for designing networks of marine reserves', *Science*, vol 298, no 5600, pp1991–1993

Schaefer, M. B. (1954) 'Some aspects of the dynamics of populations important to the management of the commercial marine fisheries', *Bulletin of the Inter-American Tropical Tuna Commission*, vol 1, pp27–56

Schnute, J. T. and Haigh, R. (2006) 'Reference points and management strategies: Lessons from quantum mechanics', *ICES Journal of Marine Science*, vol 63, no 1, pp4–11

Schnute, J. T., Maunder, M. N. and Ianelli, J. N. (2007) 'Designing tools to evaluate fishery management strategies: Can the scientific community deliver?', *ICES Journal of Marine Science*, vol 64, no 6, pp1077–1084

Schwach, V., Bailly, D., Christensen, A.-S., Delaney, A. E., Degnbol, P., van Densen, W. L. T., Holm, P., McLay, H. A., Nielsen, K. N., Pastoors, M. A., Reeves, S. A. and Wilson, D. C. (2007) 'Policy and knowledge in fisheries management: A policy brief', *ICES Journal of Marine Science*, vol 64, no 4, pp798–803

Shelton, P. A. (2007) 'The weakening role of science in the management of groundfish off the east coast of Canada', *ICES Journal of Marine Science*, vol 64, no 4, pp723–729

Shelton, P. A. and Rivard, D. (2003) *Developing a Precautionary Approach to Fisheries Management in Canada – the Decade following the Cod Collapses*, NAFO Scientific Council Research Document 03.1

Shepherd, J. (1991) 'Report of the special session on management under uncertainties', *NAFO Scientific Council Studies*, vol 16, pp59–77

Shin, Y.-J. and Shannon, L. J. (2010) 'Using indicators for evaluating, comparing and communicating the ecological status of exploited marine ecosystems: The IndiSeas project', *ICES Journal of Marine Science*, vol 67, no 4, pp686–691

Shin, Y.-J., Rochet, M.-J., Jennings, S., Field, J. G. and Gislason, H. (2005) 'Using size-based indicators to evaluate the ecosystem effects of fishing', *ICES Journal of Marine Science*, vol 62, no 3, pp384–396

Shin, Y.-J., Shannon, L. J., Bundy, A., Coll, M., Aydin, K., Bez, N., Blanchard, J. L., Borges, M. D. F., Diallo, I., Diaz, E., Heymans, J. J., Hill, L., Johannesen, E., Jouffre, D., Kifani, S., Labrosse, P., Link, J.S., Mackintosh, S., Masski, H., Moomann, C., Neira, S., Ojaveer, H., Abdallahi, K. O. M., Perry, I., Thiao, D., Yemane, D. and Cury, P. M. (2010a) 'Using indicators for evaluating, comparing, and communicating the ecological status of exploited marine ecosystems: Setting the scene', *ICES Journal of Marine Science*, vol 67, no 4, pp692–716

Shin, Y.-J., Bundy, A., Shannon, L. J., Simier, M., Coll, M., Fulton, E. A., Link, J. S., Jouffre, D., Ojaveer, H., Mackintosh, S., Heymans, J. J. and Raid, T. (2010b) 'Can simple be useful and reliable? Using ecological indicators to represent and compare the states of marine ecosystems', *ICES Journal of Marine Science*, vol 67, no 4, pp717–731

Smith A. D. M., Fulton, E. J., Hobday, A. J., Smith, D. C. and Shoulder, P. (2007) 'Scientific tools to support the practical implementation of ecosystem-based fisheries management', *ICES Journal of Marine Science*, vol 64, no 4, pp633–639

Smith, A. D. M., Sainsbury, K. J. and Stevens, R. A. (1999) 'Implementing effective fisheries-management systems: Management strategy evaluation and the Australian partnership approach', *ICES Journal of Marine Science*, vol 56, no 6, pp967–979

Smith, S. J., Hunt, J. J. and Rivard, D. (eds) (1993) *Risk Evaluation and Biological Reference Points for Fisheries Management*, Canadian Special Publication of Fisheries and Aquatic Sciences

Stanbury, K. B. and Starr, R. M. (1999) 'Applications of geographic information systems (GIS) to habitat assessment and marine resource management', *Oceanologica Acta*, vol 22, no 6, pp699–703

Stefansson, G. (2003) 'Multi-species and ecosystem models in a management context', in M. Sinclair and G. Valdimarsson (eds) *Responsible Fisheries in the Marine Ecosystem*, FAO and CABI Publishing, Wallingford, Oxford, pp171–188

Stefansson, G. and Rosenberg, A. A. (2006) 'Designing marine protected areas for migrating fish stocks', *Journal of Fish Biology*, vol 69, pp66–78

Stelzenmüller, V., Rogers, S. I. and Mills, C. M. (2008) 'Spatio-temporal patterns of fishing pressure on UK marine landscapes, and their implications for spatial planning and management', *ICES Journal of Marine Science*, vol 65, no 6, pp1081–1091

Stenseth, N. C. and Rouyer, T. (2008) 'Destabilized fish stocks', *Nature*, vol 452, no 7189, pp825–826

Stergiou, K., Christou, E.D., Georgopolilos, D., Zenetos, A. and Souvermezoglou, C. (1997) 'The Hellenic seas: Physics, chemistry, biology and fisheries', *Oceanography and Marine Biology: An Annual Review*, vol 35, pp415–538

Stokes, K., Butterworth, D. S., Stephenson, R. L. and Payne, A. I. L. (1999) 'Confronting uncertainty in the evaluation and implementation of fisheries-management systems: Introduction', *ICES Journal of Marine Science*, vol 56, no 6, pp795–796

Su, Y. (2000) 'A user-friendly marine GIS for multi-dimensional visualization', in D. Wright and D. Bartlett (eds) *Marine and Coastal Geographic Information Systems*, Taylor & Francis, London

Symes, D. (2007) 'Fisheries management and institutional reform: A European perspective', *ICES Journal of Marine Science*, vol 64, no 4, pp779–785

Triantafyllou, G., Petihakis, G. and Allen, I. J. (2003) 'Assessing the performance of the Cretan Sea ecosystem model with the use of high frequency M3A buoy data set', *Annales Geophysicae*, vol 21, pp365–375

Tusseau, M. H., Lancelot, C., Martin, J.-M. and Tassin, B. (1997) '1D coupled physical-biological model of the north-western Mediterranean Sea', *Deep-Sea Research II*, vol 44, pp851–880

Tyldesley, D. (2006) 'A vision for marine spatial planning', *Ecos*, vol 27, pp33–39

UNEP and IOC/UNESCO (United Nations Environment Programme and Intergovernmental Oceanographic Commission/United Nations Educational, Scientific and Cultural Organization) (2009) *An Assessment of Assessments, Findings of the Group of Experts: Start-up Phase of a Regular Process for Global Reporting and Assessment of the State of the Marine Environment Including Socio-economic Aspects*, UNEP and IOC-UNESCO, Progress Press Ltd, Valetta, Malta

UNESCO (2006) A *Handbook for Measuring the Progress and Outcomes of Integrated Coastal and Ocean Management*, Intergovernmental Oceanographic Commission Manuals and Guides 46, ICAM Dossier 2, Paris

Valavanis, V. D. (2002) *Geographic Information Systems in Oceanography and Fisheries*, Taylor & Francis, London

Valavanis, V. D., Georgakarakos, S., Koutsoubas, D., Arvanitidis, C. and Haralabous, J. (2002) 'Development of a marine information system for cephalopod fisheries in eastern Mediterranean', *Bulletin of Marine Science*, vol 71, no 2, pp867–882

Van Houtan, K. and Pauly, D. (2007) 'Snapshot: Ghost of destruction', *Nature*, vol 447, pp123–124

Varma, H. (2000) 'Applying spatio-temporal concepts to correlative data analysis', in D. Wright and D. Bartlett (eds) *Marine and Coastal Geographic Information Systems*, Taylor & Francis, London

Walters, C. and Maguire, J.-J. (1996) 'Lessons for stock assessment from the northern cod collapse', *Reviews in Fish Biology and Fisheries*, vol 6, no 2, pp125–137

Walters, C., Pauly. D. and Christensen, V. (1999) 'Ecospace: Prediction of mesoscale spatial patterns in trophic relationships of exploited ecosystems, with emphasis on the impacts of marine protected areas', *Ecosystems*, vol 2, no 6, pp539–554

Watson, R., Alder, J. and Walters, C. (2000) 'A dynamic mass-balance model for marine protected areas', *Fish and Fisheries*, vol 1, no1, pp94–98

Chapter 6

The Ecosystem Approach to Marine Planning and Management – The Road Ahead

Sue Kidd, Andrew J. Plater and Chris Frid

This chapter aims to:

- Summarize the main findings that emerge from the earlier chapters;
- Discuss overall conclusions from the seminar series related to natural science, social science and policy and governance perspectives; and
- Identify future research priorities that could assist the more effective delivery of EA in marine planning and management.

Introduction

This volume consolidates the lessons learnt from a series of seminars organized to bring natural scientists, social scientists and the policy and practice community together to consider the challenges of applying the ecosystem approach (EA) to marine planning and management. During the course of the seminar series the world changed; new science revealed yet more about the scale of threats to the marine ecosystem; new international accords were signed; the Copenhagen conference on climate change failed to address issues of carbon dioxide emissions, so entraining further warming and acidification of the oceans; and new legislative measures were introduced, including the UK's 2009 Marine and Coastal Access Act, which incorporates provision for the development of a marine spatial planning framework and the designation of networks of marine protected areas. It could be thought that this shifting scene would detract from the currency of the arguments presented here, however, in many respects, the opposite is true as it serves to illustrate the complexity of factors influencing effective planning and management of the marine environment and the need for the holistic and transdisciplinary responses that are envisaged in EA and are the focus of discussion here. In this final chapter we synthesize the contributions set out in the previous chapters and the general conclusions of the seminar series, and reflect on the way forward.

EA as a guide for managing human/natural environment relationships

A central theme throughout this volume has been the close relationship between human societies and the sea – the world's greatest common region covering 71 per cent of the Earth's surface. Since prehistoric times the marine environment has been used as a critical provider of food and resources and as a means of waste disposal (Desse and Desse-Berset, 1993; Jackson, 2001; Jackson et al, 2001). Today, modern uses of the ocean, such as transport, mineral extraction, pipelines and cables, and offshore wind energy installations, have all increased the range and footprint of human activities. Looking to the future, continued global population growth (rising from around 6 billion in 2010 to an anticipated peak of around 9 billion in 2050), combined with international efforts to combat climate change and ongoing technological development, can be anticipated to only intensify humanity's relationship with the sea and dependence upon the wide array of ecosystem services it provides. As we have seen, this relationship is complex and, in addition to delivering many human benefits, human activity is at present compromising the ecological integrity of the marine environment upon which all life on Earth ultimately depends. The considerable challenge this situation presents was revealed in the Millennium Ecosystem Assessment (MEA) that concluded that marine and coastal ecosystems are being used unsustainably and are deteriorating faster than other ecosystems (UNEP, 2006). It is within this context that a new sense of urgency has emerged to develop more coherent and effective marine planning and management arrangements in many parts of the world.

EA is the overarching paradigm that is guiding this activity. With antecedents stretching back to the 1970s and before, work related to the implementation of the UN Convention on Biological Diversity (CBD) has played a key role in fleshing out the conceptual underpinnings of EA. Particularly significant here has been the development of a series of guiding principles and supporting operational guidance which build out from a core understanding that humans are an integral component of ecosystems and that human activity is the main focus of planning and management intervention. From this starting point, EA provides a useful conceptual framework for structuring marine planning and management arrangements. This places considerable emphasis on the development of a strong human dimension including close and ongoing engagement with all relevant sectors of society at all points in the decision-making process and with the economic context in which planning and management activities take place. Equally, it sets out a series of parameters that reflect the complexities of ecosystem functioning that should be respected in developing marine planning and management regimes and responses. These include the need to: recognize interaction between ecosystems; focus on structure and functioning; recognize varying temporal scales and lag effects; and, perhaps most importantly of all, recognize that change is inevitable, understanding is partial and adaptive approaches to planning and management are needed if we are to make progress.

While the breadth of the ecosystem approach for integration, transdisciplinary exchange and discourse has considerable benefits, it also suffers from being all things to all people and it would be wrong to suggest that its application is easy or straightforward. Practitioners in the ecosystem approach have emerged across different stakeholder and decision- and policy-making communities who often actually have very different ideas about what it entails, what it aims to achieve and how it is operationalized. The example given in Chapter 5 illustrates this with the distinction between an EA to fisheries and ecosystem-based fisheries management. Although it may seem semantic, the basis for decision-making and the identification of management goals is quite different between these two perspectives which are guided and informed by different authorities and expert panels: ecosystem-based management prioritizes the needs of the ecosystem first, while EA stresses that objectives are a matter for societal choice and in reality this tends to mean modification of established policy directions to more effectively accommodate ecosystem drivers and impacts (as discussed in Chapter 3). While the latter approach may seem inconsistent with precautionary principles, a key argument supporting EA is that enforcement of controls over human activities in order to protect ecosystem health is problematic without wider societal 'buy in'. Other arguments supporting this view are presented in Chapter 2. However, the difficulties in applying EA do not end here. The connectedness and nested scales (in time and space) of ecosystems also makes the definition of planning and management boundaries in line with EA ambitions problematic. For example, the optimal design of a Marine Protected Area (MPA) has to incorporate factors and organisms that have a wide variety of operational scales. Furthermore, consideration has to be given to a web of social and economic factors, from the local scale right up to the international level, that also influence the use of the designated area – which may, of course, vary through the year and over the long term. Beyond such technical concerns there is also a community of sceptics, rooted in traditional disciplines, sectors and knowledges, who believe that EA offers nothing new and regard it merely as a fad that muddies the water and presents a threat to clear decision-making informed by sound science or economic imperatives and to the future of much needed research and monitoring or new development.

Now enshrined in international and national laws and policy statements, including those related to the United Nations Convention on the Law of the Sea (UNCLOS), it is therefore not surprising that attention is focusing on the need to develop a clearer understanding of the implications of EA and how its active implementation can be encouraged. The seminar series discussions reflected in Chapter 1 highlighted three key areas where further attention would be beneficial. The first relates to the human dimension of EA. Here, reflecting the points set out above, it is recognized that the terminology of EA is itself problematic as it is open to a number of quite different interpretations. Critically, it does not necessarily convey the importance of placing humans at the centre of planning and management attention as we cannot even begin to contemplate managing the complexities of the wider marine ecosystem. Changes in terminology have been discussed but dismissed on the grounds of causing further confusion. What is needed instead is ongoing education about the holistic ambitions of EA among natural scientists and environmental managers and in the community at large, for it is a concept that is relevant to all areas of human planning not just those directly related to the natural environment. A key message for those engaged in planning and management of the sea however is that ongoing and active engagement of all sections of society in decision-making is critical.

A second key area identified for further deliberation relates to the information needed to support sound decision-making. The availability of good quantitative and qualitative data related to both the natural and human components of marine systems is clearly important and presents particular challenges in the marine context where information availability is much more limited than on land, but good data is insufficient on its own. What is also required is imaginative translation of data so that decision-makers can have as clear an understanding as possible of where the economic, social and environmental costs and benefits of different planning and management responses may lie. This in turn requires understanding of spatial and temporal dynamics and the structure and functioning of marine ecosystems which are all key areas of research effort at the present time and where ongoing funding and support is needed. However, the complexity and uncertainty that is inherent in marine (and indeed other) ecosystems means that in reality decisions have to be taken with less than perfect knowledge. EA promotion of adaptive approaches to planning and management is helpful in making it clear that action need not and indeed cannot wait until the perfect data set or model has been produced.

Thirdly, Chapter 1 stresses the importance of connecting marine planning and management understanding and concerns to wider agendas. This includes closer integration of marine and terrestrial planning as the latter is of critical importance in determining the levels of human pressure on the sea. In this respect, the development of a strong EA-informed marine planning and management perspective could play a key part in debates about future human development trajectories and provide additional insights as to the appropriateness of the currently dominant understanding that continuing economic growth moderated by ecological modernization is the way ahead. To date such discussions have been dominated by land-based interests that are often ignorant of the impacts and potential consequences of continuing economic and population growth on the largest and arguably most significant ecosystem on Earth – the sea. A view from the sea may suggest an alternative conclusion.

Developing the human dimension of EA: Connecting to spatial planning for the land

It is inconceivable that activities on land would take place in an unregulated 'free for all' and so every nation state has developed some form of control over development via a terrestrial planning system. It therefore seems logical that, as pressures on the marine environment grow, a close connection to spatial planning for the land may also be of value in developing the human dimension of marine planning and management (see Chapter 2). It is encouraging that there is already evidence that, particularly in relation to matters of process, marine planning and management is in fact now beginning to draw quite heavily upon terrestrial planning practice and in particular on the current spatial planning paradigm. There is, however, perhaps a danger of adopting this in an unquestioning and simplistic way without due recognition of key differences between terrestrial and marine environments, the history behind the development of spatial planning and the sophistication of the concept which draws together strands from a very long history of planning thought.

In terms of elaborating on the human dimension of EA with the benefit of marine planning in mind, a review of terrestrial planning experience reveals how views on the purpose and process of any planning activity are likely to evolve over time and serves to illustrate the significance of EA Principle 1, which indicates that the objectives of planning and management of the sea are not value-free but are, in fact, matters of societal choice. This suggests that active debate and deliberation in order to achieve clarity and some degree of public consensus about the desired attributes to be strived for is important. In turn, reflections on terrestrial planning experience underline the significance of EA Principle 12 concerning the engagement of all sections of society (including the scientific community) in this process and how this should inform the skill set available within marine planning and management teams and the process through which decisions are made. At various phases of history, terrestrial planners have been seen as designers, scientists and as communicators and mediators between different interests, with the current spatial planning paradigm envisaging a combination of all these. What does this suggest for the training needs of the emerging profession of marine planning and management? This line of thinking also reveals the relevance of EA Principle 11, which encourages the consideration of all forms of relevant information, including scientific and indigenous and local knowledge in decision-making. Perhaps most significantly though, it again draws attention to EA Principle 9, which encourages recognition that change is inevitable and indicates that this relates as much to the human as well as the natural dimensions of marine ecosystems. Interestingly, this highlights the need to look beyond simply charting changing patterns of human use of the sea to considering the underlying cultural, legal, administrative, political, social and economic norms and practices that determine them. In this context, marine planning needs to give due attention to anticipating the shifts that may take place and, in line with EA Principle 6, set plans and objectives for the long term. It is also perhaps critically important here to recognize that marine spatial planning (like planning for the land) is not simply about identifying and describing the factors of change and using them as inputs to decision-making in a mechanistic way, but that it also entails a creative dimension that should not be ignored or disparaged. With the anticipated growth in global population over the first part of the 21st century, both sound science and imagination, as well as full engagement with the human dimension of EA will be needed in order to deliver more sustainable patterns of development in our seas.

Integrating terrestrial and marine planning and management agendas

The difficulties in charting a more sustainable relationship between humans and the sea cannot be underestimated and the complexities of factors linking terrestrial and marine agendas are illustrated very well by the discussion in Chapter 3 related to experience in Europe. Here it is shown, for example, that continued over-exploitation of marine biological resources has been accompanied by continued deterioration of marine ecosystems in European seas and that the EU's Common Fisheries Policy (CFP) 2002–2012, which provides a key mechanism with which to balance economic, social and environmental considerations, has failed to stem this decline. However, a radical new approach to the CFP is emerging for the period after 2012 which seems to offer better prospects for European seas. The consolidation and continuing development of EU directives related to environmental regulation (such as the Marine Strategy

Directive) also seems to support this view, as does the advent of a number of marine spatial planning initiatives and legislation, in which The Netherlands, Belgium, Germany and the UK are leading protagonists.

In parallel, the Maastricht, Nice and Lisbon treaties mark a period of continuing constitutional development, aimed at strengthening and integrating EU institutions and governance and this also seems to be supportive of EA ambitions related to integration. However, this stops far short of the kind of integration that is necessary to convert the present monetary union into a stable arrangement where environmental matters truly achieve the same level of consideration as economic matters. While the Lisbon and Gothenburg Strategies represent an important first step towards developing more environmentally benign long-term goals for EU development, the emphasis is still very much upon economic growth and jobs and an ecological modernization world view. The current deep recession in the EU is unlikely to challenge this at least in the short term. Interestingly, the main initiatives aimed at more integrated planning and management for marine areas in Europe are located largely outside the 'core' of the EU, where human pressures on the sea are of an order of magnitude greater than the vast marine periphery. It is in these waters in particular where integrated approaches to terrestrial and marine planning would be most beneficial.

Ecosystem goods and services

Turning to the information challenges of EA, Chapter 4 indicated how the ecosystem goods and services concept offers the potential to make marine ecosystems tangible to the non-specialist. It provides a strong transdisciplinary framework for knowledge exchange and the development of shared learning across different disciplines, sectors and stakeholders who may have very different knowledges, perspectives, priorities and values that are not easily conveyed between actors in the decision-making process. Recognizing that healthy, functioning marine ecosystems provide goods and services that underpin economic sustainability, well-being and health – even regulating climate, cycling key elements, nutrients, pollutants, etc. in the environment, and making us more resilient to extreme events (floods, droughts) – provides a common ground for integrated planning and management of the sea where EA is at the heart of decision-making. Nurturing the ongoing delivery of ecosystem goods and services can perhaps be developed here as a shared over arching goal. For example, the sustained provision of fish for food and other uses (fertilizer, medicines, recreation, and so on) is a goal that all stakeholders can identify, relate to and work towards, whatever their background or expertise – yet at the root of this is the healthy, functioning marine ecosystem.

While EA acknowledges the importance of understanding ecosystems, their structure, their function, etc., the planning framework it provides does not get bogged down with issues such as 'insufficient data' or 'not understanding the full complexity of the system'. The science community will always rely on (and perhaps hide behind) further study and research to provide insight and understanding for the formulation of models, as well as data gathering to provide evidence of trends, distributions or for model validation. This 'better understanding' would not necessarily lead to an operational mechanism for marine planning and management – although it would probably use all the available budget and resources! Focusing on the goods and services that ecosystems 'provide', being careful not to frame ecosystems as being at the beck-and-call of humans, takes us beyond the need for comprehensive understanding of ecosystems. In essence, the ecosystem structural components, their interrelationships and functions are understood to a sufficient level so that management action can be taken to prevent any further decline in ecosystem services and to perhaps facilitate the conditions that enhance them or promote recovery. Within EA, the goods and services concept also potentially provides us with the following: practical reasons for why action must be taken now despite imperfect understanding, else we lose or at least see a serious decline in ecosystem goods and services (for example, collapse of large marine ecosystems, extirpation of species, loss of food or degradation of water quality); the goals for planning and management, to ensure that these goods and services are prioritized in the decision-making process and thus maintained; and a means by which to track and measure management actions that can then be both precautionary and adaptive, in other words, is the management regime having the intended effect or are there emerging undesirable consequences?

More contentious is the mechanism by which ecosystem goods and services are placed within the planning process. The Millennium Ecosystem Assessment (MEA) framework provides a clear illustration of how human health and well-being, and indeed economic sustainability, are intimately linked to ecosystem services that play very different roles. Different disciplines, communities and stakeholders will clearly place different values on these services and, of course, prioritize them accordingly. How are these diverse perspectives and values then brought together in an integrated planning framework for decision-making? One approach that has sought to provide a unified currency for considering the diversity of demands made on the marine ecosystem and a way of evaluating their impacts has been the development of economic valuation techniques and their application to the full gambit of goods and services drawn from the marine ecosystem. Costanza et al (1997) valued the contribution of marine ecosystems to the global economy at around US\$21 billion per annum.

Some fields of public policy development have seen economic valuation as a key mechanism for being able to judge the relative costs and benefits of different management strategies, and this technique can be readily applied to tangible ecosystem goods and services to which a monetary value can be assigned. The tonnage of fish catches clearly has a price per unit catch and, hence, an economic value if catches can be managed at a given level whether it be artisanal or large-scale industrial, or even as subsistence. This principle can be readily applied to other 'provisioning' ecosystem services, as well as many 'regulating' services, such as the economic cost of losses due to flooding, loss of potable water or changing patterns of temperature and rainfall. However, less tangible 'supporting' and, in particular, 'cultural' ecosystem services are not so readily reduced to an economic value. Indeed, many arguments stress that it is crucial that they are not. Reducing aesthetic beauty or spiritual well-being to an economic cost means that these aspects can then be 'outbid' by other factors that may jeopardize these characteristics in favour of enhanced total economic value. Hence, in advocating the tremendous transdisciplinarity of ecosystem services as a tool for knowledge exchange, this does not necessarily extend to the economic valuation of these services as the basis for decision-making.

Data and modelling tools and the role of science in EA

From our review of data, models and tools for implementing an EA to managing the marine environment (Chapter 5), it is clear that many lessons can be learned from fisheries management in terms of principles and practical steps. A considerable body of research has been undertaken on the use of indicators with which to identify the impact of fishing on the marine environment and, hence, provide operational tools for tracking or measuring the success of a management regime. Indeed, there is a long history of monitoring and the use of resources devoted to the acquisition of monitoring data, upon which the effectiveness of indicators as a management assessment tool is based. The founding principles of these indicators are entirely aligned with those of marine spatial planning (Ehler and Douvere, 2009). One of the key issues is that ecosystems are not readily encapsulated in metrics that can then be used to measure the success (or failure) of a management regime or particular action. Fundamentally, indicators are not yet able to fully capture key aspects of ecosystem dynamics and function, feedbacks and linkages, buffering, sensitivity, and so on. Is this a fundamental problem? No doubt the ecology and conservation communities feel that it is, especially if they take the view that understanding ecosystem status, interaction, function, resilience and contingency are at the heart of defining management goals. Does the emphasis on ecosystem goods and services avoid this problem? Probably not because the indicators and metrics with which to track and measure the success of any management regime – in other words, mean length and lifespan, trophic level of landings, and so on – are not readily related to specific aspects of ecosystem dynamics or function. There is no doubt that there are numerous methods for tracking and measuring progress (data, monitoring networks, models and approaches for building management scenarios and options, areas for closure, and so on) but there is some debate as to whether these actually map onto defined management goals. If they do not, then they serve no purpose.

Ecosystem models are still in their infancy, particularly when it comes to complex ecosystem models that aim to predict change as a result of internal dynamics or external forcing. Hence, they should be viewed as tools for identifying the direction of change in response to changing climate, environment and/or human action. Models are often viewed with caution by stakeholders and decision-makers because of their scientific focus (in other words, not related to policy objectives), their intangibility (poorly communicated to the non-specialist) and because many do not incorporate socio-economic data. Hence, trust in their output and the interpretation of data needs to be built with stakeholders through dialogue and clear communication.

The widely held view that management action should be precautionary and adaptive is an important principle if the knowledge of ecosystems is incomplete, as with, for example, alternative stable states. The use of appropriate indicators and metrics is, therefore, of critical importance if action is to be taken soon enough to prevent inadvertent ecosystem changes that threaten valuable goods and services, or indeed lead to significant ecosystem degradation. The precautionary principle is even more difficult to implement when (i) natural ecosystem variability and (ii) shifting baselines due to climate change will also be expressed in the data. Good knowledge of intrinsic ecosystem dynamics and the likely consequences of external forcing (climate change, ocean acidification, eutrophication of coastal waters), both of which have the capacity to be modelled and thus tested using scenarios, provides the evidence that underpins adaptive management. Pragmatically, are the resources and capacity available to ensure that the monitoring required to support adaptive management can be delivered? Undoubtedly, this will vary from place to place across the world's oceans. All participants in the governance of the marine environment need to embrace the concept of adaptive management based on a process of ongoing 'experimentation', using the best but inevitably imperfect understanding, tracked by monitoring and delivered in the context of periodic policy adjustment.

In terms of what we can learn from fisheries management with respect to marine planning and management, perhaps two areas can be highlighted. The first is the use of current practice that has evolved through iteration and has, thus, brought us to a process that includes management options underpinned by evidence and robust mechanisms for evaluation. This extends to the use of techniques for multi-layer data handling and visualization and ecosystem models that incorporate fisher behaviour and socio-economic considerations. These tools are utilized to support decision-making and to scope alternative management regimes. It is, therefore, easy to see how these could be developed further to incorporate further sectoral considerations and policy requirements. The second area is that of communication and knowledge exchange, particularly in relation to uncertainty in the evidence base, the nature and timing of responses to the evidence and the politicization of the process (Rice, 2005). As mentioned previously in the context of including short-term social and economic considerations and long-term ecological priorities in the decision-making process, our consideration of past and present practice shows there is a clear need, will and established means to ensure that the ecological community acknowledges the costs inherent in its preferred management goals, and engages with the social and economic sciences to co-produce strategies for how these costs can be addressed. This is even more critical when we consider the potential breadth of considerations and perspectives in integrated marine planning and management.

Overall conclusions from the seminars

General

EA in marine planning and management is now embodied in international law and directives. Much has been done to elaborate what EA entails and understanding is developing. The time has come for us to shift the focus of debate from conceptualizing EA to issues related to its implementation.

Natural science related

Our knowledge of marine ecosystem dynamics is incomplete and we need to accept that we have to make progress with less than perfect data and recognize that expert judgment is an essential part of marine planning and management activity. The EA focus on adaptive management informed by ongoing monitoring and underpinned by knowledge, data and modelling scenarios is key in the marine context.

Experimentation with marine planning and management tools and models helps us to appreciate both the extent and limitations of our knowledge and also priorities for improved understanding and data gathering.

The ecosystems services concept is an effective mechanism for transdisciplinary knowledge sharing and for overcoming the relative powers of different knowledges that delay action and hamper progress. While it places the ecosystem as the root of human health, well-being and economic sustainability, care must be taken to ensure that ecosystems are not regarded as providing outputs, such as goods and services, simply for the good of humankind.

Natural scientists need clearly defined objectives and goals for marine planning and management to guide their work. These tend to lack clear definition and detail at the present time. EA emphasizes that objectives and goals are matters for societal choice but these need to be informed by conversations with natural scientists.

Capacity building is required among natural scientists in EA implementation. Many currently charged with implementation, mainly through the acquisition and interpretation of monitoring data, are rooted in the traditions and research frontiers of ecosystem science or have become constrained by established practice in fisheries management. It is easy to underestimate the challenge that these scientists face, especially in terms of policy developments, new technologies and in tackling the wealth of sectoral information needs in MSP.

Social science related

There is a critical need to better understand the social/human aspects of marine ecosystems as this is the key area that planning and management activity can influence.

It is important that we do not just equate social/human considerations with economics (for example, economic valuation of ecosystem goods and services).

EA emphasizes the importance of close engagement with stakeholders, communities and the wider political system, however, in many cases, stakeholder relationships with the sea are indirect and understanding of 'a sense of place' in the sea is ill-developed. Similarly, an awareness of long-term ecosystem considerations has to be built to counter short-term economic priorities. This awareness/knowledge gap needs to be addressed to encourage more meaningful policy dialogue and informed definition of policy objectives and goals.

A tiered approach to the definition of objectives seems desirable with higher level/national objectives informed by local level stakeholder as well as natural science understanding. There is a need to recognize that values underpinning objectives will change over time and vary between scales as well as individual/ sectoral perspectives, so mechanisms for ongoing information exchange and debate will have an important part to play.

In developing objectives and goals, more horizon-scanning work would be beneficial, recognizing that the marine environment is more of a blank canvas than the land and there may therefore be an opportunity to be more visionary in approach. In this context, food and energy security and low carbon development are all likely to be key future policy drivers.

Policy/governance related

We must not underestimate the difficulties in achieving more integrated planning and management of the sea. Significant cultural change is needed for this to be realized. Joined-up policy approaches that integrate sectors, tiers of governance and terrestrial and marine areas should be key features of EA for marine areas; however, the legacy of 'silo working' is still a major barrier in most marine situations. Despite this, there is evidence that the slow process of structural change seems to be beginning.

Another key issue in the implementation of an EA to marine planning and management is the disparity in understanding, capacity and progress across the globe. We have identified a significant spatial variation in the extent to which different countries and different sectors within these countries have adopted EA. While some nations lead in the development of an EA, such as fisheries management in Canada, others merely deliver monitoring as a statutory requirement. This presents significant problems for those nations that are required to meet international guidance but have limited resources or capacity to do so, or indeed nation states about to join the EU who are going through an accelerated phase of activity to meet policy needs. Capacity and resources, building knowledge and, indeed, shared knowledge, gathering data on appropriate metrics and acting in a timely fashion are therefore significant challenges to implementation.

In integrating planning and management activity across the land/sea interface, it will be useful to recognize that the sustainable resource management underpinning of emerging marine planning and management activity provides a very different starting point from that of long-established landward planning regimes, which as yet tend to have a strong economic and social as opposed to environmental emphasis.

Political accountability in marine planning and management decision-making needs further consideration and development.

Future research priorities

Governance of the marine environment is identified as a key, cross-cutting theme for future research. Useful areas of investigation include: issues related to sectoral, territorial and organizational integration, both within marine areas and from land to sea; alternative forms of partnership working; and the promotion of culture change in support of EA-based marine planning and management activity.

Developing a better understanding of what an adaptive planning and management approach to the sea might entail is a second key area for further work, particularly how to develop and use ecosystem indicators in tracking and measuring progress towards management goals. In this context, ongoing review and reflection on emerging practice experience is considered critical at this point in time. Key issues here might include comparison of matters of purpose and content as well as matters of process, such as boundary definition, planning and management time horizons, monitoring and review arrangements, and how to evaluate the success of management strategies against a moving baseline of climate-induced environmental change.

Good data are needed to underpin planning and management activity, but data availability, compatibility and accessibility are major issues in the marine environment. Further work is needed to: gain a more robust understanding of the key environmental, social and economic data that are needed to support decision-making; develop indicators that can be linked to ecosystem properties and dynamics; explore how quantitative and qualitative data and different sources of knowledge can be harnessed to best effect; consider the case for varying the frequency, resolution and scale at which different data sets might be collected; and consider data management protocols that might facilitate connectivity between different tiers of marine governance. Further development of related models and tools could also inform understanding here.

Further research helping to define the purpose of marine planning and management would also be beneficial, this might usefully include horizon-scanning initiatives as well as exploration of issues associated with the definition of good ecological status.

Innovative ways of engaging with stakeholders and politicians to develop a more refined 'sense of place' in the sea and to take a longer term sustainability perspective would assist the delivery of new marine planning and management responsibilities. Further exploration of the potential of virtual reality technology and other education and engagement tools and techniques would also be helpful here.

Finally, reflecting on the above, further cross-disciplinary, multidisciplinary, transdisciplinary and, indeed, intra-disciplinary initiatives targeted at informing the delivery of EA in the marine environment are needed. There is a continuing case for drawing together not only natural and social scientists and marine planning and management practitioners but also a wider array of disciplinary inputs including economists, statisticians with expertise in uncertainty and risk analysis, historians, artists and creative writers, cultural anthropologists and psychologists, for all of these perspectives and others have a part to play in informing future developments. For example, it is now more than 100 years since international efforts began to manage fisheries. In that time, the natural and social science aspects of fisheries have been developed but it really is only in the last few decades that debate, argument and finally consensus have emerged. Now, 110 years after the first international agreements were signed, fisheries management is beginning to take account of the role of fish in the ecosystem, the effects of the ecosystem on fish and the importance of social and economic structures. We must hope that the painful lessons of the development of fisheries management over the past century, together with the equally painful lessons that can be drawn from the parallel evolution of planning for the land, will help to inform the delivery of EA in the new era of marine planning and management that is emerging.

References

Costanza, R., d'Arge, R., de Groot, R., Farber, S., Grasso, M., Hannon, B., Limburg, K., Naeem, S., O'Neill, R. V., Paruelo, J., Raskin, R. G., Sutton, P. and van den Belt, M. (1997) 'The value of the world's ecosystem services and natural capital', *Nature*, vol 387, pp253–260

Desse, J. and Desse-Berset, N. (1993) 'Pè che et en Mèditerranèe: Le temoinage des os,' in J. Desse and F. Audoin-Rouzeau (eds) *Exploration des Animaux Sauvages à Travers le Temps*, Editions APDCA, Juan-les-Pins, pp327–339

Ehler, C. and Douvere, F. (2009) *Marine Spatial Planning: A Step-by-step Approach Toward Ecosystem-based Management*, Intergovernmental Oceanographic Commission and Man and the Biosphere Programme, IOC Manual and Guides No 53, ICAM Dossier No 6, UNESCO, Paris

Jackson, J. B. C. (2001) 'What was natural in the coastal oceans?', *Proceedings of the National Academy of Sciences of the United States of America*, vol 98, pp5411–5418

Jackson, J. B. C., Kirby, M. X., Berger, W. H., Bjorndal, K. A., Botsford, L. W., Bourque, B. J., Bradbury, R. H., Cooke, R., Erlandson, J., Estes, J. A., Hughes, T. P., Kidwell, S., Lange, C. B., Lenihan, H. S., Pandolfi, J. M., Peterson, C. H., Steneck, R. S., Tegner, M. J. and Warner, R. R. (2001) 'Historical overfishing and the recent collapse of coastal economy', *Science*, vol 293, pp629–638

Rice, J. C. (2005) 'Every which way but up: The sad story of Atlantic groundfish, featuring Northern Cod and North Sea Cod', *Bulletin of Marine Science*, vol 78, no 3, pp429–465

UNEP (United Nations Environment Programme) (2006) *Marine and Coastal Ecosystems and Human Wellbeing: A Synthesis Report Based on the Findings of the Millennium Ecosystem Assessment*, UNEP, Nairobi, 76p

Index

Page numbers in *italic* refer to Figures, Tables and Boxes. EA represents 'ecosystem approach'.